Springer Series in Optical Sciences   Volume 9

Edited by David L. MacAdam

Springer Series in Optical Sciences

Edited by David L. MacAdam

High-Power Lasers and Applications

Proceedings of the Fourth Colloquium on Electronic
Transition Lasers in Munich, June 20–22, 1977

Editors
K.-L. Kompa and H. Walther

With 142 Figures

Springer-Verlag Berlin Heidelberg GmbH 1978

Dr. KARL LUDWIG KOMPA

Max-Planck-Gesellschaft zur Förderung der Wissenschaften e.V.
Projektgruppe für Laserforschung
D-8046 Garching bei München, Fed. Rep. of Germany

Professor Dr. HERBERT WALTHER

Max-Planck-Gesellschaft zur Förderung der Wissenschaften e.V.
Projektgruppe für Laserforschung und
Sektion Physik der Universität München
D-8046 Garching bei München, Fed. Rep. of Germany

Dr. DAVID L. MACADAM

68 Hammond Street, Rochester, NY 14615, USA

ISBN 978-3-662-15400-7   ISBN 978-3-540-35942-5 (eBook)
DOI 10.1007/978-3-540-35942-5

© by Springer-Verlag Berlin Heidelberg 1978
Originally published by Springer-Verlag Berlin Heidelberg New York in 1978.

Softcover reprint of the hardcover 1st edition 1978

2153/3130-543210

Preface

The High-Power Lasers and Applications Conference was held in Munich, June 20 - 22, 1977. The conference took place simultaneously with the "Laser 77, International Congress and Trade Fair" at the Munich Fair Ground. The meeting was a continuation of a series of colloquia on electronic transition lasers previously held in the United States. The main topics of the conference were: high-power VUV, UV, visible and IR lasers, including an analysis of laser systems, technology and laser concepts. Also, some applications to nonlinear optics, chemical kinetics and spectroscopy, particularly with respect to isotope separation, were discussed.

The conference was attended by 95 scientists representing Austria, Canada, England, Finland, Germany (FRG), Germany (GDR), France, Israel, Italy, The Netherlands, and the U.S.A.

The organizers acknowledge financial support from the Deutsche Forschungsgemeinschaft, the U.S. Air Force Office of Scientific Research, the U.S. Air Force European Office of Aerospace Research and Development (EOARD) and the U.S. Army European Research Office, as well as from the companies Coherent Radiation, Spectra Physics and Cryophysics.

Furthermore, we thank our colleagues Dr. Steven N. Suchard and Professor Jeffrey I. Steinfeld for coordinating the U.S. contribution to the conference. We are grateful to Frau Maischberger for administrative assistance.

Karl Ludwig Kompa
Herbert Walther

December 1977

Contents

Part III *Other Laser Systems*

Part IV *Frequency Conversion*

Part V *Applications*

Part I

Excimer Lasers

Excimer Lasers

Ch.A. Brau

University of California, Los Alamos Scientific Laboratory
Los Alamos, NM 87545, USA

A number of excimer lasers have now been demonstrated spanning the spectrum from green to the vacuum ultraviolet, and others have been proposed.
Although the different classes of excimer lasers have rather different
properties, making them useful for different purposes, the rare gas
halides have demonstrated the highest power and efficiency, and promise to
be the most widely useful. Overall efficiency in excess of 1% and pulses
as large as 350 J have been achieved from krypton fluoride, and higher
efficiency and much larger pulses are predicted. A wide variety of uv
wavelengths is available using stimulated Raman scattering.

Excimers are molecules which are bound in an electronically excited state
but not in the ground state, and emit characteristic broad band, bound-free
radiation. See Fig. 1. The bound-free nature of the laser transition assures

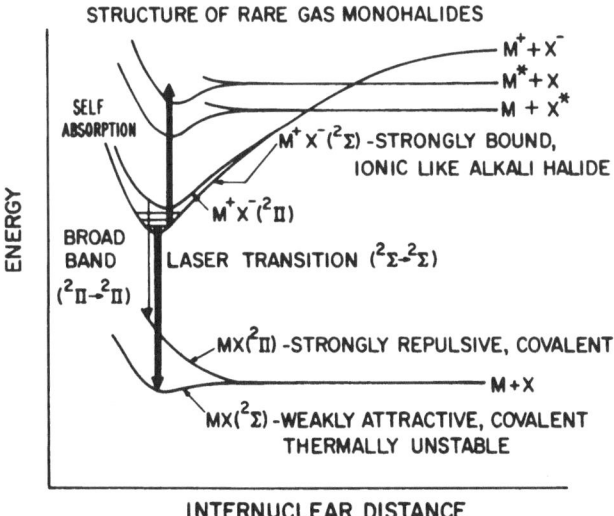

STRUCTURE OF RARE GAS MONOHALIDES

Fig. 1 Potential energy diagram of rare gas halide molecule showing laser and
self-absorption transitions.

an inversion and prevents bottlenecking in the lower laser level. However, while the broad band nature of the transition offers some tunability, it also lowers the gain, causing excimer lasers to have high threshold pumping powers. Several classes of excimer lasers have now been demonstrated, including the rare gases, the rare gas oxides, the rare gas monohalides, and mercury (gain only). Similar lasers, in which the lower laser level is bound but in which lasing takes place to a point near dissociation limit, include hydrogen, sodium, the halogens and the recently discovered mercury monohalides. As summarized in Table I, the wavelengths of excimer lasers span the spectrum from green to the vacuum ultraviolet. Other excimer lasers have been proposed and are actively being investigated, including complexes of mercury and the rare gases with various metal atoms. However, the most successful excimer lasers have been the rare gas monohalides, which are the only excimer lasers now available commercially. These lasers have demonstrated high efficiency and high pulse energies previously available only at infrared wavelengths. However, they have their difficulties and limitations as well. Because of their current importance, the remainder of this talk is addressed to the rare gas monohalides.

Table 1. Wavelengths of Excimer Lasers

Species	Wavelength (nm)	Reference
Xe_2	172	Koehler, et al., Appl. Phys. Lett. 21, 198 (1972)
Kr_2	146	Hoff, et al., Appl. Phys. Lett. 23, 245 (1973)
Ar_2	126	Hughes, et al., Appl. Phys. Lett. 24, 488 (1974)
XeO	540	Powell, et al., Appl. Phys. Lett. 25, 730 (1974)
KrO	558	Powell, et al., Appl. Phys. Lett. 25, 730 (1974)
ArO	558	Hughes, et al., Appl. Phys. Lett. 28, 81 (1976)
XeBr	282	Searles and Hart, Appl. Phys. Lett. 27, 435 (1975)
XeF	353	Brau and Ewing, Appl. Phys. Lett. 27, 435 (1975)
XeCl	308	Ewing and Brau, Appl. Phys. Lett. 27, 350 (1975)
KrF	249	Ewing and Brau, Appl. Phys. Lett. 27, 350 (1975)
ArF	193	Hoffman, et al., Appl. Phys. Lett. 28, 350 (1976)
KrCl	222	Murray and Powell, Appl. Phys. Lett. 29, 252 (1976)
Hg_2	335	Schlie, et al., Appl. Phys. Lett. 28, 393 (1976)

To better understand the performance of the rare gas halides, a few words about the spectroscopy and kinetics of these systems are in order. The laser transitions are strongly allowed, with lifetimes of the order of 10 ns and gain cross sections in excess of 10^{-16} cm^2. However, there are also many photoabsorbers in the ultraviolet, including the halogen molecules and a number of transient species such as excited rare gas atoms and excimers, rare gas dimer ions, halogen negative ions, and even the excited rare gas halide molecules themselves.

The basic reason for the higher power and efficiency of the rare gas halide lasers rests in very favorable kinetics for the formation of the upper laser level. The result of electrical excitation of rare gas mixtures is to produce excited rare gas atoms, ions, and electrons. These form rare gas halides in reactions such as

$$Kr^* + F_2 \rightarrow KrF + F \tag{1}$$

and

$$Kr^+ + F^- + Ar \rightarrow KrF^* + Ar. \tag{2}$$

Both these processes are very rapid, occurring in times typically of the order of 10 ns. In many cases these reactions are essentially 100% efficient in producing excited rare gas halide products. However, the excited products are quenched by rapid processes as well, such as

$$KrF^* + F_2 \rightarrow Kr + 3F, \tag{3}$$

and

$$KrF^* + Kr + Ar \rightarrow Kr_2F^* + Ar. \tag{4}$$

To achieve the high pumping rates needed to reach threshold, several techniques have been used, including high intensity, relativistic electron beams, fast transverse discharges, and electron beam stabilized discharges. In the future direct nuclear pumping may be accomplished using a pulsed nuclear reactor to initiate fission reactions in the laser. When electron beams are used to pump the laser, the relativistic electrons form excited rare gas atoms and ions, with an average energy $\bar{W} \sim 20$ eV deposited in the gas for each ion or excited state produced. Thus, in a KrF laser with a photon energy $h\nu$, the maximum possible efficiency is $\eta(\max) = h\nu$ (KrF)/$\bar{W}$ (argon) = 25%. Intrinsic efficiencies (laser energy $\div$ deposited in gas) as high as 15% have been observed from KrF [1], but overall ("wall plug") efficiencies have been much lower [2]. Using a large area electron beam, pulses as large as 350 J have been obtained from KrF [3]. Larger pulses are possible, but ultimately the pulse energy becomes limited to something of the order of a kilojoule per meter of laser length by problems of closure of the electron beam gun and self pinching of the electron beam. The overall length of the laser is limited by photoabsorption processes but pulses as large as kilojoules appear possible. The pulse repetition frequency and, therefore, the average power are limited by the rate at which the foil (through which the electron beam enters the laser) can be cooled and the laser medium cleaned up by flow and acoustic damping.

Discharges are potentially more efficient than electron beams since it takes less energy to form an excited rare gas atom with a thermal electron than it takes to form an ion with a relativistic electron. Except in helium [4], more than 80% of the discharge energy can appear as electronic excitation of the rare gas [5]. See Fig. 2. The maximum possible efficiency is limited, of course, by the ratio of the photon energy $h\nu$ to the rare gas excitation energy E^*. Thus, in a KrF laser $\eta(\max) = h\nu(KrF)/E^*(Kr) = 50\%$. In practice, much lower efficiencies are observed for a number of reasons. Discharges in electronegative gases are fundamentally unstable and if they are not stabilized externally the impedance rapidly collapses, after breakdown, and spatial nonuniformities develop. Thus, such discharges cannot be operated at optimum conditions. Nevertheless, simple, low inductance, fast transverse discharges have produced pulses larger than 700 mJ, with a pulse length of the order of 20 ns and an overall efficiency (laser energy $\div$ stored energy) in excess of 1% [6]. A pulse repetition frequency of 100 Hz has been demonstrated in a separate experiment [7]. Increasing the pulse repetition frequency by more than an order of magnitude appears possible, but will depend on advances in switching technology. Larger pulses can be achieved by using electron beam ionization of the gas to stabilize the discharge [8]. Pulses as large as 50 J and as long as 300 ns have been achieved this way [3]. However, the electric field and, therefore, the gain are limited in such discharges, making efficient extraction of the laser energy difficult. See Fig. 2.

6

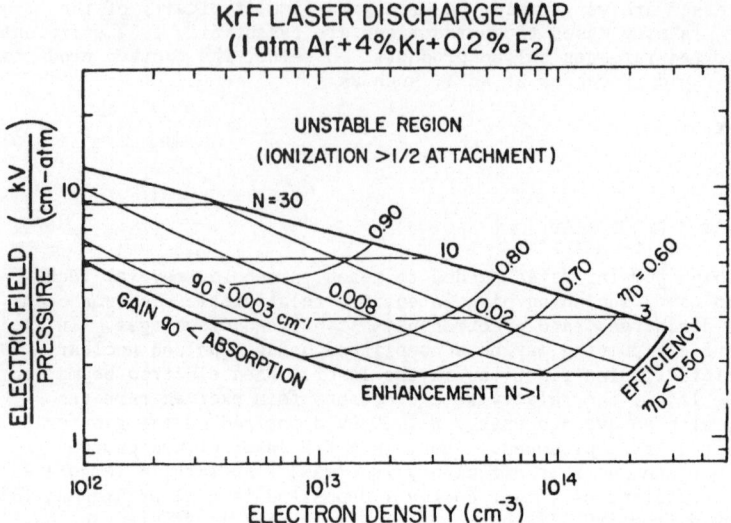

Fig. 2 Operating map for discharge-pumped krypton fluoride laser using Ar with 4% Kr and 0.2% F_2 at 100 kPa total pressure

The pulse length and wavelength of the rare gas halide lasers can be modified in a number of ways. Pulses as long as 1 µs have been achieved with electron beam pumping [9], but pulses longer than 10 µs become limited by consumption of the molecular halogen in the gas and the growth of medium inhomogenieties. Short pulses can be obtained by mode-locking, with a minimum pulse length limited by the duration of the gain pulse [10]. Shorter pulses can be amplified within the available bandwidth, but the short lifetime and high gain of the rare gas halides prevent them from storing large amount of energy to be extracted in a single large pulse. When large, short pulses are needed, the rare gas halides may be useful as efficient, powerful pumps for other lasers, such as the iodine photodissociation laser [12], or it may be possible to use Raman pulse compression techniques [13]. Wavelength shifting may also be accomplished in a number of ways. XeF (351 nm) lasers can be used to pump visible laser dyes, and KrF (248 nm) lasers have been used to pump dyes for the near uv region [14]. It should be possible to achieve at least discrete tunability deeper into the uv by pumping other, simpler molecules. Stimulated Raman scattering has been used to shift XeF (351 nm) to 585 nm in barium vapor [15], and a multitude of wavelengths throughout the uv have been achieved by Raman scattering KrF and ArF radiation from H_2, D_2, CH_4 and liquid N_2 [16] (see Fig. 3). In the future, much shorter wavelengths may be possible by tripling (4-wave mixing) the rare gas halide lasers, as has been done with Xe_2 lasers [17].

The power and efficiency of the rare gas halide lasers should make them useful for a variety of applications. For chemical processing, rare gas halide laser photons are available for only pennies per mole (for the electricity) and their use for separating oxygen [18] and hydrogen [19] isotopes

SRS-Generated UV Wavelengths

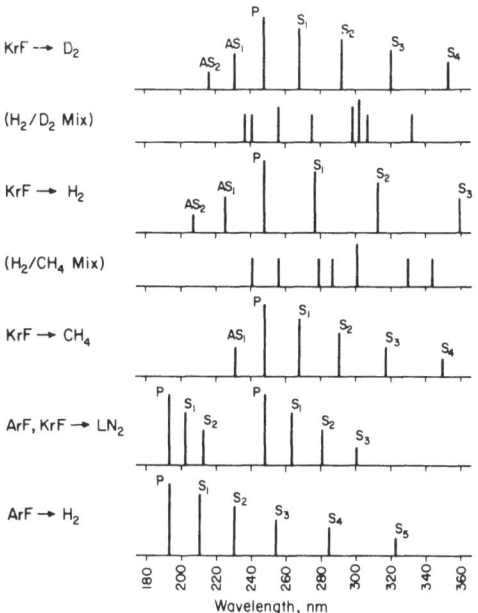

Fig. 3 Wavelengths achieved by stimulated Raman scattering of ArF and KrF lasers.

has already been discussed. In the future they may be useful for micro-machining, lithography, deep space communications, remote sensing of the atmosphere, optical radar, or anywhere that intense, cheap uv photons are needed.

1   C.A. Brau and J.J. Ewing in Electron Transition Lasers, ed. J.I. Stein-feld, 195-198 (MIT Press, Cambridge, MA, 1976).

2   M.L. Bhaumik, R.S. Bradford and E.R. Ault, Appl. Phys. Lett., 28, 23, (1976).

3   R. Hunter, 7th Winter Colloquium on High Power Visible Lasers, Park City, Utah, Feb. 16-18, 1977.

4   A.E. Greene, C.A. Brau, T.R. Loree, R.C. Sze, and S.D. Rockwood, 5th Conference on Chemical and Molecular Lasers, St. Louis, MO, April 18-20, 1977.

5   J.H. Jacob and J.A. Mangano, Appl. Phys. Lett., 28, 724 (1976).

6   W.J. Serjeant, private communication.

7   B. Goddard and M. Vannier, Opt. Commun., 18, 206 (1976).

8   J.D. Daugherty, J.A. Mangano, and J.H. Jacob, Appl. Phys. Lett., 28, 581 (1976).

9   L.F. Champagne, J.G. Eden, N.W. Harris, N. Djeu, and S.K. Searles, Appl. Phys. Lett., 30, 160 (1977).

10 C.P. Christensen, L.W. Braverman, W.H. Steier, and C. Wittig, Appl. Phys. Lett., 29, 424-425 (1976).

11 I.V. Tomov, R. Fedosejevs, M.C. Richardson, W.J. Serjeant, A.J. Alcock, and K.E. Leopold, Appl. Phys. Lett., 30, 146 (1977).

12 J.C. Swingle, C.E. Turner, Jr., J.R. Murray, E.V. George, and W.F. Krupke, Appl. Phys. Lett., 28, 387 (1976).

13 J.R. Murray and A. Szöke, 7th Winter Colloquium on High Power Visible Lasers, Park City, Utah, Feb. 16-18, 1977.

14 D.G. Sutton, and G.A. Capelle, Appl. Phys. Lett., 29, 563 (1976).

15 N. Djeu and R. Burnham, Appl. Phys. Lett., 30, 473 (1977).

16 T.R. Loree, R.C. Sze, and D.L. Barker, Appl. Phys. Lett., 30 in press, (1977).

17 M.H.R. Hutchinson, C.C. Ling, and D.J. Bradley, Opt. Commun., 18, 203 (1976).

18 R.K. Sander, T.R. Loree, S.D. Rockwood, and S.M. Freund, Appl. Phys. Lett., 38, 150 (1977).

19 J.B. Marling, Chem. Phys. Lett., 34, 84 (1975).

Coherent Radiation Generation
at Short Wavelengths VUV and XUV Pulses

D.J. Bradley

Optics Section, Blackett Laboratory, Imperial College
London SW7 2BZ, England

Abstract

Recent developments of tunable-frequency rare-gas excimer and exciplex
oscillators have extended the range of laser radiation into the vacuum ultraviolet region.   By third- and higher-order resonantly enhanced nonlinear
processes, coherent radiation can be generated at wavelengths as short as
35nm (XUV).   Applications of these new pulsed sources of short-wavelength
coherent radiation are considered.

1. Introduction

The development of short-wavelength lasers had to await the arrival of
sufficiently intense pumping sources since the laser pumping power required
scales, at least, as the fourth power of the lasing frequency.   The years
1970 and 1971 saw the introduction of high-voltage electron-beam technology
for pumping the molecular Xenon vacuum ultraviolet (VUV) laser (1) and the
electron-beam controlled discharge excited atmospheric-pressure CO_2 laser (2).
Earlier the invention of the nitrogen laser (3) operating at 337nm had
encouraged the development of the rapid discharge technology required for
shorter-wavelength lasers.   Together with dye lasers, the N_2 laser had
shown the capability of molecular electronic transitions for laser action, as
first proposed by HOUTERMANS in 1960 (4).   The first VUV laser, operating on
the Lyman band of H_2, was realised in 1970 by fast-discharge high-voltage
pumping (5) and was soon followed by lasing in the Weiner band at 116nm (6).
This is still the shortest wavelength at which a laser oscillator has worked
to date.   With the extensive developments in the 1960's of high-current
electron-beam sources, the technology was available at the beginning of the
present decade for an attack on the problem of producing efficient tunable-
frequency lasers in the ultraviolet (UV) and VUV spectral regions.   Largely
due to the pioneering work of Martin and his colleagues at the AWRE Aldermaston
Laboratory, electron-beams with kiloampere currents and energies in the
megawatt range were available (7) to pump high-pressure gases rapidly enough
to both overcome the short fluorescence lifetimes of the upper laser levels
and to create noble gas excimers (4) in sufficient densities and volumes to
provide adequate gain for broad-band VUV laser action.   Thus in 1972 high-
pressure molecular lasers held great promise for the efficient production
of tunable, coherent radiation over the VUV spectral region(i) and this
promise has since been realised.

(i) UV = 200nm to 400nm; VUV = 100nm to 200nm; XUV = 10nm to 100nm

It is now as easy to produce narrow-band, frequency-tunable low beam-divergence megawatt laser pulses in the VUV (12,13) as with dye lasers operating in the visible, (Table 1).

Table 1 Typical properties of Xenon excimer laser

	Untuned	Tuned
Gas pressure	7 - 10 kTORR	7 - 10 kTORR
Pulse energy	15 - 55 mJ	2 - 20 mJ
Pulse duration	3 - 16 ns	3 - 16 ns
Peak power	~ 5 MW	~1 MW
Beam divergence	1 - 4 mR	1 - 3 mR
Tuning range	-	170 - 176 nm
Bandwidth	~10 Å	~1.5 Å (0.15 Å at 0.2 - 0.3 MW)

Electron-beam excited noble gases are also seen as the most likely pumping sources, via either photolytic pumping or excitation transfer schemes, for new high-power laser systems for fusion studies (14,15).

Within the past year rapid developments of high-efficiency UV lasers (16) employing rare gas monohalides and pumped directly by e-beams, by e-beam sustained discharges and by rapid discharges have made available a wide range of coherent sources throughout the VUV and UV spectral regions (see Table 2). Frequency tuning in the UV over a range of $3000 cm^{-1}$ (17) and active mode-locking (18) have already been achieved.   VUV operation has recently been obtained with ArCl (19) and F_2 (20). It is likely that the halogens may not be unique in forming excimers and exciplexes with noble gases. HUTCHINSON (21) has proposed that, by analogy with the alkali metals, noble gas hydroxide exciplexes would be expected to be produced in collisions with water or hydrogen peroxide molecules.   Initial experiments (22) have produced strong broad-band fluorescence at ~ 224 nm from a mixture of Xenon, Argon and water vapour excited by high-energy electrons.

Also in the past year coherent radiation has been produced in the XUV spectral region by generation of the third harmonic frequency (57nm) of the Xe_2 laser tuned into resonance with a two-photon transition of the nonlinear medium.   Shortly afterwards, using a similar experimental arrangement the 20th (53.2nm) and 28th (38nm) harmonics of picosecond pulses from a mode-locked Nd:YAG laser (25,26) were generated in Helium and Neon.   By fifth order mixing of the 266nm (second-harmonic) 532nm (fourth harmonic) and the fundamental 1.06μm radiation a range of longer wavelengths in the XUV has also been produced.   In this approach, the nonlinear mixing efficiency is enhanced by the use of high-powers and of near-resonances with multiple-photon transitions of the mixing gas, and by tight-focussing.   The use of a tunable-frequency

Table 2 Noble gas excimer and exciplex lasers

Excimer	Exciplex	Emission Wavelength (nm)	Mode of Excitation		
			E-Beam	E-Beam Stabilised Discharge	Rapid Discharge
Xe_2		172	X		
Kr_2		146	X		
Ar_2		126	X		
	XeF	354	X	X	X
	XeCl	308	X	X	X
	XeBr	282	X		
	KrF	249	X	X	X
	KrCl	222	X		X
	ArF	193	X		X
	ArCl	175			X

narrow-band VUV laser permits optimization by tuning to exact resonance with a two- or three- photon transition, in addition to using tight-focussing conditions. This method has the incidental advantage of requiring lower laser powers, which helps to avoid the onset of level shifting, and breakdown due to the optical-frequency electric fields (25,26). However for systematic studies of higher-order resonant nonlinear processes in the VUV and XUV spectral regions a repetition-rate laser system is obviously desirable, particularly for ease of optimization of the experimental parameters. We have been constructing such a laser in the Blackett Laboratory for these and other purposes.

2. High Repetition-rate Tunable VUV Laser

To achieve the high-efficiencies potentially obtainable with relativistic electron-beam pumping of noble gas excimer and exciplex lasers, it is necessary to achieve good coupling of the electron-beam source to the high-pressure laser medium. The coaxial diode pumping arrangement (28,29) has permitted both a high-degree of uniformity in the excitation of the laser gas with the deposition of ~50% of the electron beam energy (13,23,29). While large scale versions of the coaxial diode system are under study for laser fusion applications (30,31) we have concentrated our efforts on developing a high-repetition rate system. Our earlier Xe_2 lasers (23,29) were restricted to operation at a rate of, at most, one or two pulses per minute due to the limitations of the high-voltage pulsers employed (Hewlett Packard Febetron Types 706 and 705). Thus it was essential to design and construct a new

high-voltage power supply.   The equivalent circuit of the LARK pulser
(32) developed in collaboration with AWRE, Aldermaston, is shown in Fig. 1.

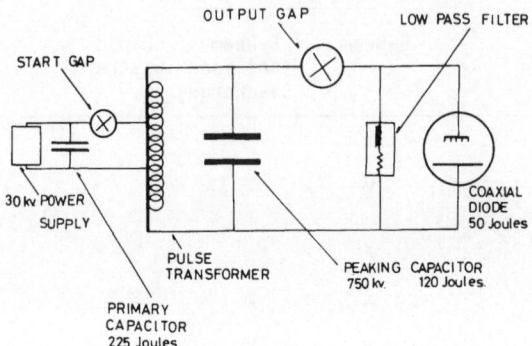

Fig. 1   Equivalent circuit diagram of 750 kV e-beam pulser
for repetition-rate operation

This arrangement, which can generate 1 GW pulses (750kV, 50J, 50ns) at a
repetition rate of 10Hz, occupies a volume of dimensions 2m x 1.25m x 1.25m.
Compared with Marx-bank systems the use of an auto-transformer gives the
advantage that only two switches are required, with a consequent reduction
in jitter to ∼1%.   The coaxial diode load has three symmetrically arranged
razor-blade emitters of 20cm length (Fig. 2) instead of the perforated
titanium sheet used in previous systems.   The pulser has been operated at
7Hz for periods of hours without any noticeable change in current or voltage
waveforms.

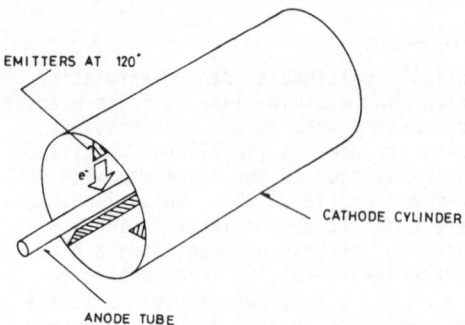

Fig. 2   Xe$_2$ laser coaxial diode configuration

Figure 3 shows the complete laser system with some of the lead shielding removed for photography. (Even for the small X-ray dose produced per shot, operation at $\sim$10,000 pulses/hour requires properly designed shielding). The Xenon gas circulation system can be seen mounted on top of the laser diode. The instantaneous rise in temperature of the gas after pumping is $\sim$ 700°C and the anode tube temperature increases by $\sim$ 30°C after a few seconds, by thermalization with the gas. Since absorption by ground-state Xenon molecules increases rapidly with increasing temperature (33) gas heating strongly affects the laser intensity and beam-divergence, so rapid circulation through a heat-exchanger is needed. Output energy is also dependent upon

Fig. 3 Photograph of repetition-rate tunable-frequency Xe_2 VUV laser. E-beam pulser is in box at rear

the Xenon pressure since the upper level is populated by kinetic processes involving three body collisions (34,35). However while the output energy increases with increasing gas pressure, at higher pressures the range of the electrons becomes smaller than the anode tube radius, so the centre of the gas is no longer effectively excited. In Fig. 4 each experimental point

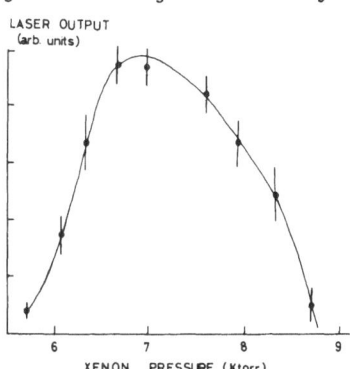

LASER OUTPUT
(arb. units)

XENON PRESSURE (Ktorr.)

Fig. 4 Variation of coaxial diode Xe_2 laser output energy with gas pressure

is the average of 10 laser pulses, showing the high degree of reproducibility of the laser output energy, particularly at the optimum working pressure of ∼ 7kTorr.   Peak laser power is limited to 2MW by damage to the laser optical components and most of the evaluation measurements have been carried out with resonator mirror reflectivities of 65% and 20%.   Frequency narrowing and tuniı is produced by an intracavity quartz prism as in earlier systems (Fig. 5)

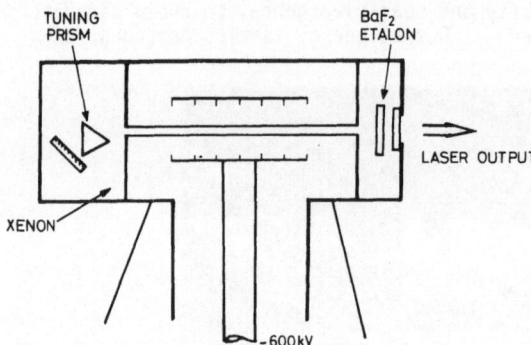

Fig. 5 Optical arrangements of Xe_2 laser showing tuning prism resonator mirrors

3. Coherent XUV Radiation Generation

Experiments are currently underway in our laboratory to generate in Helium the fifth harmonic frequency (35nm) of the Xe_2 laser, by tuning to a three-photon resonance ($1s^2$ - 1s2p) corresponding to an initial laser wavelength of 175.3nm, (Fig. 6).

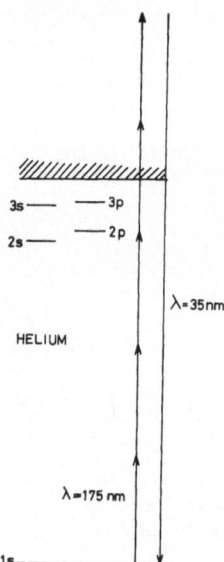

Fig. 6 Partial energy diagram of He showing the $1s^2$ - 1s2p three-photon resonance for fifth-harmonic conversion from 175.3nm to 35nm

The experimental arrangement is the same as that used for third-harmonic generation in Argon (Fig. 7). The laser beam is tightly focussed by a BaF$_2$ lens into the mixing cell, half of which also acts as a photo-ionization counter of the XUV photons. The two halves of the gas cell are separated by a thin aluminium foil which filters out the fundamental radiation so as to avoid photoemission from the photoionization gauge electrode.

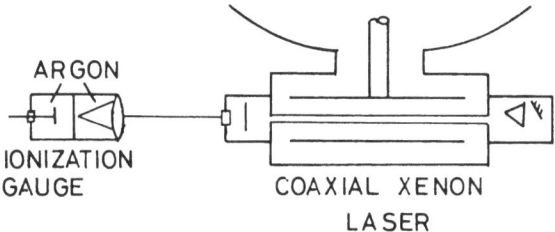

ARGON

IONIZATION
GAUGE   COAXIAL XENON
                              LASER

Fig. 7 Experimental arrangement for generation and detection of third-harmonic radiation at 57nm

As the Xenon laser frequency is tuned through the two-photon resonance ($3p^6$ - $3p^55p$) in Argon a resonance curve centred at 170.9nm is traced out (Fig. 8), showing how the third-order susceptibility in Ar is dominated by

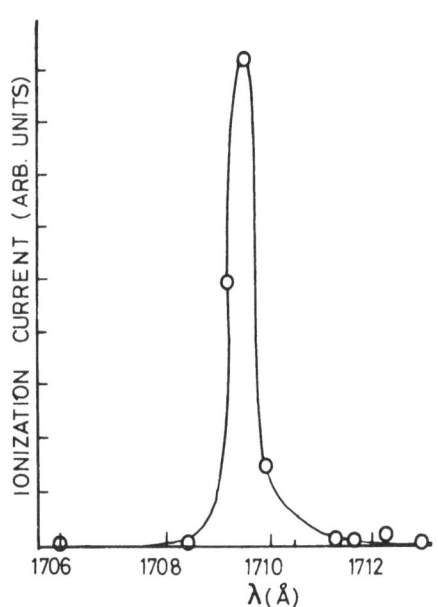

Fig. 8 Third-harmonic generated energy as a function of the Xe$_2$ laser wavelength showing the two-photon resonance at 170.9nm

the two-photon resonance. The halfwidth of $\sim$ 0.1nm of the resonance curve is equal to the operating bandwidth of the laser.

Discussion and Conclusion

It is clear that the availability of coherent radiation at XUV and VUV wavelengths will have a major impact on physical and biological sciences. In conjunction with the excitation of molecular vibrational structures by infrared lasers, VUV lasers can be expected to play an increasingly important role in photochemistry, particularly in studies of molecular photoionization. Other likely applications include phase-contrast and holographic microscopy at VUV and XUV wavelengths, selective detection of complex molecules, and the determination of molecular structures. Some of these applications will require picosecond and sub-picosecond pulse excitation to achieve stop-motion recording. Fortunately the broad bandwidths available in excimer and exciplex lasers permit the generation of very short duration pulses. It is interesting to note that the generation of ultra-short pulses in the visible and infra-red regions is now limited by the optical carrier-wave frequency (36,37). It is thus necessary to go to shorter wavelengths to generate pulses shorter than 100 femtoseconds (10^{-13}s). As can be seen from Table 3, to generate a pulse of duration 10 femtoseconds containing 50 optical cycles, would require a maximum carrier wavelength of 60nm.

Table 3 Relationship between laser pulse duration and optical carrier-wave frequency.

Pulse duration Δt	Wavelength of optical period = Δt	Maximum carrier wavelength for pulse of 50 optical cycles
10^{-9}s	30 cm	6 mm
10^{-12}s	300 μm	6 μm
10^{-14}s	3 μm	60 nm

Other scientific applications could be VUV photoelectron spectroscopy of solids, and for the diagnosis of high-temperature, high-density plasmas in laser fusion research using short wavelength picosecond pulses. VUV lasers (particularly at the Lyman-α wavelength) would also be useful for the diagnosis of magnetically confined fusion plasmas. More practical applications are for lithography, for the production of electronic microcircuit elements and for materials processing exploiting the advantages of lower reflectivity and deeper penetration at shorter wavelengths. Finally it is likely that the most practicable approach to soft X-ray laser systems will be to use nonlinear processes to up-convert the frequencies of longer-wavelength lasers in the VUV and UV for amplification in plasmas, where population inversion has been achieved by recombination, charge-transfer or photolytic pumping processes. The recent extension of streak-camera technology into the XUV and X-ray spectral regions (36) to provide a time-resolution of $\sim$ 5ps will greatly facilitate experimental investigations, since radiative lifetimes are in the picosecond range at these wavelengths.

Acknowledgements

The author wishes to thank Dr. M. H. R. Hutchinson, Dr. C. C. Ling, Mr. C. B. Edwards and other members of the Blackett Laboratory Optics Section Laser Group whose work is described in this paper. The enthusiastic cooperation and help of Mr. J. C. Martin and his colleagues at AWRE Aldermaston Laboratory in the design and construction of the LARK pulser system is gratefully acknowledged. Financial support has been obtained from the Science Research Council and the UKAEA Culham Laboratory.

References

1. N. G. Basov, V. A. Danliychev and Yu. M. Popov: JETP Lett. 12, 329 (1970)

2. C. A. Fenstermacher, M. J. Nutter, J. P. Rink and K. Boyer: Bull. Amer. Phys. Soc., 16, 42 (1971)

3. H. G. Heard: Nature, 200, 667 (1963)

4. F. G. Houtermans: Helv. Phys. Acta., 33, 933 (1960)

5. R. T. Hodgson: Phys. Rev. Lett., 25, 494 (1970); R. W. Waynant, J. D. Shipman, R. C. Elton and A. W. Ali: Appl. Phys. Lett., 17, 383 (1970)

6. R. W. Waynant: Phys. Rev. Lett., 28, 533 (1972); R. T. Hodgson and R. W. Dreyfus: Phys. Rev. Lett., 28, 536 (1972)

7. H. H. Fleischmann: Physics Today, 28, 35 (1975)

8. H. A. Koehler, L. J. Ferderber, R. L. Redhead and P. J. Ebert: Appl. Phys. Lett., 21, 198 (1972)

9. A. Wayne Johnson and J. B. Gerardo: J. Appl. Phys., 45, 867 (1974)

10. P. W. Hoff, J. C. Swingle and C. K. Rhodes: Opt. Commun., 8, 128 (1973); Appl. Phys. Lett., 23, 245 (1973)

11. W. M. Hughes, J. Shannon and R. Hunter: Appl. Phys. Lett., 24, 488 (1974)

12. D. J. Bradley, D. R. Hull, M. H. R. Hutchinson and M. W. McGeoch: Opt. Commun., 14, 1 (1975)

13. E. G. Arthurs, D. J. Bradley, C. B. Edwards, S. D. Domanski, D. R. Hull, C. C. Ling and M. H. R. Hutchinson: In Proc. Intern. Conf. on Electron Beam Research and Technology, ed. by G. Yonas, (Sandia Laboratories Report SAND 76-5122, 1976. Vol. II) p. 193

14. J. R. Murray and C. K. Rhodes: Lawrence Livermore Laboratory Rept. UCRL-51455, 1973.

15. S. D. Rockwood: Los Alamos Scientific Laboratory Rept. LA-UR-73-1031, 1973

16. J. J. Ewing and C. A. Brau: In Springer series in Optical Sciences Vol. 3: "Tunable Lasers and Applications" Ed. by A. Mooradian, T. Jaeger and P. Stokseth (Springer, Berlin, Heidelberg, New York 1976). p.21 and references therein.

17. R. Burnham: Private communication

18. C. P. Christensen, L. W. Braverman, W. H. Steier and C. Wittig: Appl. Phys. Lett., 29, 424 (1976)

19. R. W. Waynant: Appl. Phys. Lett., 30, 234 (1977)

20. J. K. Rice, A. K. Hays and J. R. Woodworth: Appl. Phys. Lett., 31, 31 (1977)

21. M. H. R. Hutchinson: UK Patent application No. 25861/77 (1977)

22. M. H. R. Hutchinson: Private communication

23. D. J. Bradley, M. H. R. Hutchinson and C. C. Ling: In Springer series in Optical Sciences Vol. 3 "Tunable Lasers and Applications" Ed. by A. Mooradian, T. Jaeger and P. Stokseth (Springer, Berlin, Heidelberg, New York, 1976) p. 41

24. M. H. R. Hutchinson, C. C. Ling and D. J. Bradley: Opt. Commun., 18, 203 (1976)

25. J. Reintjes, R. C. Eckhardt, C. Y. She, N. E. Karangelen, R. C. Elton and R. A. Andrews: Phys. Rev. Lett., 37, 1540 (1976)

26. J. Reintjes, C. Y. She, R. C. Eckhardt, R. A. Andrews and R. C. Elton: Appl. Phys. Lett., 30, 380 (1977)

27. C. Y. She and J. Reintjes: Appl. Phys. Lett., 31, 95 (1977)

28. D. J. Bradley and M. H. R. Hutchinson: UK Patent No. 14102/74 (1974)

29. D. J. Bradley , D. R. Hull, M. H. R. Hutchinson and M. W. McGeoch: Opt. Commun., 11, 335 (1974)

30. Lawrence Livermore Laboratory Laser Program Commercial Report. UCRL-50021-75 (1976) pp. 538-549

31. Sandia Laboratories Laser-Fusion Research Progress Report SAN 77-0157 (1977) pp. 69-81

32. C. B. Edwards, M. D. Hutchinson, J. C. Martin and T. H. Storr: AWRE Report SSWA/JCM/755/99 (1975)

33. D. A. Emmons: Opt. Commun., 11, 2 57 (1974)

34. A. Wayne Johnson and J. B. Gerardo: J. Chem. Phys., 59, 1738 (1973)

35. D. J. Bradley, M. H. R. Hutchinson and H. Koetser: Opt. Commun., 7, 187 (1973)

36. D. J. Bradley: In Topics in Applied Physics, Vol. 18 "Ultra-short Light Pulses, Picosecond Techniques and Applications" Ed. by S. L. Shapiro (Springer, Berlin, Heidelberg, New York 1977) p. 18

37. I. S. Ruddock and D. J. Bradley: Appl. Phys. Lett., 29, 296 (1976)

38. See e.g. ARPA/NRL X-ray laser program, NRL Memorandum Report 3482 (1977)

Dominant Formation and Quenching Processes in E-Beam Pumped ArF* and KrF* Lasers[1]

M. Rokni[2], J.H. Jacob, J.A. Mangano, J. Hsia, and A.M. Hawryluk

Avco Everett Research Laboratory, Inc.
Everett, MA 02149, USA

ABSTRACT

The dominant formation and quenching processes in E-beam pumped ArF^* and KrF^* lasers are discussed. The exciplexes are produced by irradiating Ar/F_2 and $Ar/Kr/F_2$ mixtures with a $5A/cm^2$, 150 keV E-beam. A steady state analysis is valid since the reaction times are short compared to the 300 nsec beam pulse length. The quenching of ArF^* by F_2 and Ar has been measured by analyzing the ArF^* ($B^2\Sigma_{1/2} \rightarrow X^2\Sigma_{1/2}$) fluorescence as a function of the F_2 and Ar partial pressures. We have also measured the displacement of the Ar in ArF^* by Kr to form KrF^*. The dominant quenching processes of KrF^* were identified and the rate constants were measured. The ArF^* and KrF^* are formed from the ionic states with high efficiency. Interception of the precursers can be made negligible by choosing the experimental conditions properly. The quenching of KrF^* by Ar and Kr is mainly a three body process resulting in the formation of Kr_2F^*. The emission from Kr_2F^* was observed in a broadband centered at 410 nm. We have verified that the Kr_2F^* is produced subsequent to the KrF^* formation by performing a laser saturation experiment.

I. Introduction

Much research has been performed on the rare gas monohalide exciplexes since their spectra was first reported by VELAZCO and SETSER.[1] In the two years following the VELAZCO-SETSER publication many of these molecules have been made to lase by both pure E-beam [2] and discharge pumping. [3] The most promising candidates that are scalable to high output power and efficiency are the rare gas fluorides. [4,5] To facilitate the scaling of these lasers to high average power a detailed knowledge of the kinetic processes are necessary. VELAZCO, KOLTS and SETSER [6] have

[1] This research was supported by the Advanced Research Projects Agency and monitored by the Office of Naval Research under Contract No. N00014-75-C-0062.

[2] On leave from Hebrew University of Jerusalem, Israel.

shown that the rare gas fluorides are formed with high efficiency from excited rare gases. We will show that the rare gas ions also produce these exciplex species with high efficiency. The formation rate of these excited molecules is rapid since they can be accessed through an ion channel: E-beam ionization of the rare gas followed by rapid dissociative electron attachment to the halogen and subsequent extremely rapid ion-ion recombination.

The kinetic processes were investigated by irradiating mixtures of rare gases and fluorine by a beam of fast electrons. These processes were isolated by studying the dependence of the resulting fluorescence on the partial pressure of each of the constituents of the gas mixture. By analyzing the dependence of the quasi-steady state fluorescence on the partial pressures of the rare gases and fluorine and the power deposited into the gas mixture, we obtained the various quenching rate constants. [7]

II. Experimental Set-up

The experimental set-up is shown schematically in Fig. 1, and has been described in detail elsewhere. [7] The gas mixtures were excited by a 150 keV E-beam. The mixture was contained in a teflon cell with dimensions 22 x 2 x 0.2 cm^3. The dimension of the cell along the initial E-beam direction was 0.2 cm to insure uniform energy deposition by the E-beam up to mixture pressures of 4 atm. The maximum E-beam current density was 5 A/cm^2. The current density could be further attenuated by introducing a partially transmitting screen on the high vacuum side of the foil. The E-beam pulse pulse length was 300 nsec long, enabling the fluorescence amplitude to reach a steady state. The fluorescence intensities were monitored by appropriate filters and photodiodes, and spectra were recorded on a 1 meter Hilger spectrograph. In the case of the ArF* radiation, the side of the filter facing the photodiode was coated with sodium salicylate. The sodium salicylate converts the 193 nm photons into visible and near UV radiation.

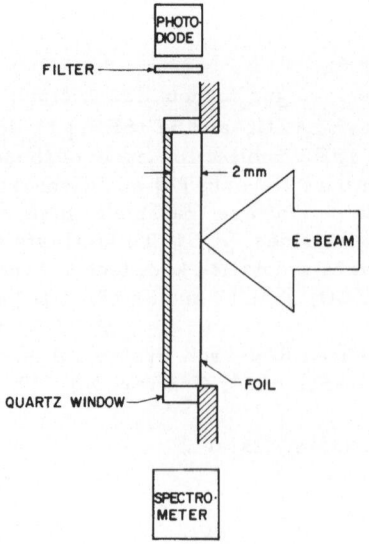

G7721

Fig. 1 Schematic of experimental set-up

III. Formation and Quenching of ArF^*

Since Ar is the main constituent in most rare gas fluoride laser mixes, most of the E-beam energy is deposited in the argon. So it is reasonable to first investigate the kinetic processes in E-beam pumped Ar/F_2 mixes.

(a) Formation of ArF^*

Table 1 lists the dominant formation kinetics for low current density (≤ 10 A/cm^2) E-beam pumped systems. About 55% of the E-beam energy deposited in the gas is channeled into Ar^+ formation as given by reaction (1). Approximately 10% of the deposited energy is channeled into Ar^* formation by the energetic secondary electrons formed in reaction (1).[8] For our experimental conditions, i. e. , Ar pressures below 4 atm, mixtures containing ≥ 2 torr of F_2 and E-beam currents ≤ 5 A/cm^2, the main loss mechanism for secondary electrons is dissociative attachment by F_2 resulting in the formation of F^-. Hence the dominant ArF^* formation mechanism proceeds via the ion channel [see reactions (4) and (7) in Table 1].

Table 1 Dominant formation kinetics for ArF^*

$\vec{e}$ + Ar	$\rightarrow$	Ar^+ + $\vec{e}$ + e_s	(1)
e_s + F_2	$\rightarrow$	F^- + F	5×10^{-9} cm^3/sec[17] (2)
e_s + Ar	$\rightarrow$	Ar^* + e_s	(3)
F^- + Ar^+ + (M)	$\rightarrow$	ArF^* + **(M)**	10^{-7} + 10^{-7} p; p < 1 atm(4)
Ar^* + F_2	$\rightarrow$	ArF^* + F	(5)
PRESSURE > 1 ATM			
Ar^+ + F^- + (M)	$\rightarrow$	ArF^* + (M)	1.1×10^{-6} cm^3/sec (4a)
Ar^+ + 2Ar	$\rightarrow$	Ar_2^+ + Ar	2.5×10^{-31} cm^6/sec[18] (6)
Ar_2^+ + F^-	$\rightarrow$	ArF^* + Ar	(7)

At pressures below an atmosphere ArF^* is mainly formed via reaction (4). In this pressure range the ion-ion equivalent two body recombination rate constant increases linearly with pressure (3-body process). This 3-body reaction becomes diffusion limited at pressures > 1 atm. Between pressures of 1-4 atm this reaction is expected to reach an effective two-body rate of $\sim 10^{-6}$ cm^3/sec. [9] At pressures ≈ 1 atm, Ar_2^+ and Ar^+ have almost the same number densities. Ar_2^+ recombines with F^- via 2-body reaction to form ArF^* (reaction 7). The molecular ions could possibly form Ar_2F^* via $Ar_2^+ + F^- + M \rightarrow Ar_2F^* + M$. However, it will be shown subsequently that this process is unimportant. Once ArF^* is formed it can radiate or be quenched by F_2 or other constituents of the gas mixture. The dominant quenching processes and measured rate constants are listed in Table 2.

Table 2 Dominant quenching processes of ArF^*

REACTION			(RATE CONSTANT)x (ArF* LIFETIME)	RATE CONSTANT[a]
$ArF^* + F_2$	$\rightarrow$	Products	$7.6 \pm 0.7 \times 10^{-18} \text{ cm}^3$	$1.9 \times 10^{-9} \text{ cm}^3/\text{sec}$
$ArF^* + Kr$	$\rightarrow$	$KrF^* + Ar$	$6.1 \pm 0.5 \times 10^{-18} \text{ cm}^3$	$1.6 \times 10^{-9} \text{ cm}^3/\text{sec}$
$ArF^* + Ar$	$\rightarrow$	Products	$3.6 \pm 1 \times 10^{-20} \text{ cm}^3$	$9 \times 10^{-12} \text{ cm}^3/\text{sec}$
$ArF^* + 2Ar$	$\rightarrow$	$Ar_2F^* + Ar$	$1.6 \pm 0.3 \times 10^{-39} \text{cm}^6$	$4 \times 10^{-31} \text{ cm}^6/\text{sec}$

[a]The rate constants have been evaluated assuming an ArF^* lifetime of 4 nsec.[10]

(b) Quenching of ArF^*

The ArF^* quenching by F_2 was determined by observing the ArF^* fluorescence keeping the Ar partial pressure fixed at 150 torr [10] and varying the F_2 partial pressure from 2-20 torr. A Stern-Volmer plot of the ArF^* fluorescence data as a function of the F_2 partial pressure is shown in Fig. 2. From this plot, the half-pressure for F_2, i.e., the pressure of F_2 where the inverse quenching rate becomes equal to ArF^* lifetime, is 4.0 torr.

To determine the quenching of ArF^* by Ar, experiments were performed keeping the partial pressure of F_2 fixed at 2 torr and varying the partial pressure of Ar from 100 torr to 4 atm. Figure 3 shows the data for a typical set of runs. Notice that the signal increases up to a pressure of about one atmosphere and then decreases slowly.

There are two possibilities for the observed decay:
(1) ArF^* quenching by Ar in two and three body processes, or
(2) Decreasing formation efficiency of ArF^*.
As the Ar pressure is increased, reaction (6) (in Table 1) occures more

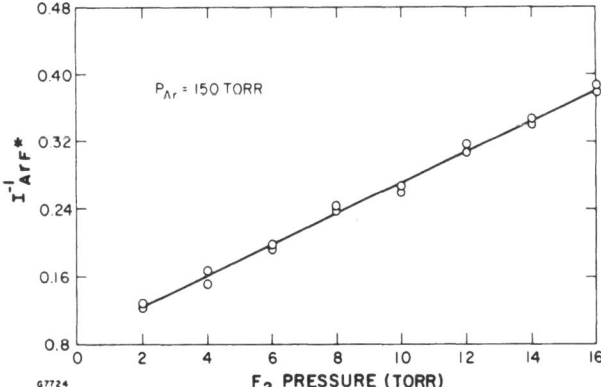

<u>Fig. 2</u> Stern-Volmer quenching curve for $ArF^* \, (^2\Sigma_{1/2})$ with F_2

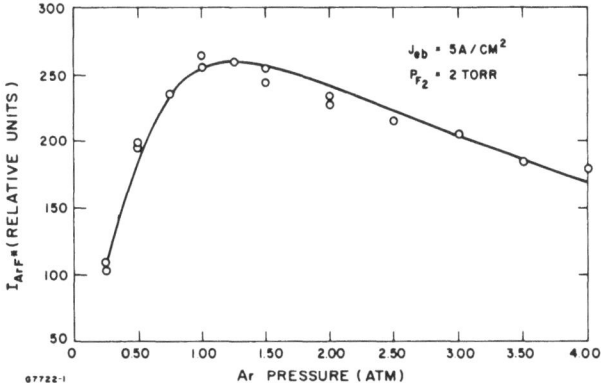

<u>Fig. 3</u> $ArF^* \, ^2\Sigma_{1/2} \rightarrow \, ^2\Sigma_{1/2}$ fluorescence in the presence of 2 torr F_2 as a function of Ar partial pressure. The points are experimental values for 5 A/cm² E-beam current. The curve is the expected ArF^* fluorescence using the measured quenching rate constants

frequently to form molecular ions Ar_2^+. In fact, at the highest pressure the density of Ar_2^+ is ten times that of Ar^+. These molecular ions will recombine with F^- and can form ArF^* or possibly the excited triatomic Ar_2F^*. The formation of the Ar_2F^* by this channel will result in a smaller formation efficiency of ArF^* and could account for the observed decrease in the fluorescence with increasing pressure. To insure that this was in fact not the case, we attenuated the E-beam current by a factor of 25. This causes a decrease of the F^- density by at least a factor of 5, resulting in a higher probability of Ar_2^+ formation at a given pressure. So changing the current factor of 25 should strongly affect the fluorescence efficiency of ArF^* if Ar_2^+ plus F^- form Ar_2F^*. Figure 4 shows the experimentally determined ratio of the ArF^* fluorescence intensity as a function of pressure when the E-beam current is changed by a factor of 25. Also shown are predicted ratios for three cases: (1) 100% of Ar_2^+ forms ArF^*, (2) 80% of Ar_2^+ forms ArF^*, and (3) 60% of Ar_2^+ forms ArF^*. From Fig. 4 we can conclude that

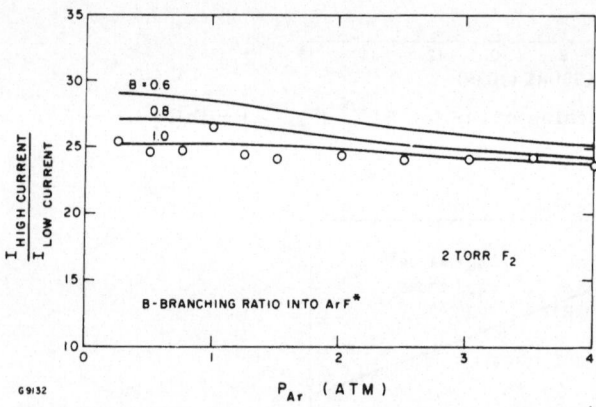

<u>Fig. 4</u> The measured and calculated ratio of ArF^* fluorescence intensities when the E-beam current is changed by a factor of 25. The points are experimental. The three curves are the calculated ratios for branching ratios of 1, 0.8 and 0.6

almost all the Ar_2^+ forms ArF^*. The results in Fig. 4 prove that the dominant product of $(Ar_2^+ + F^-)$ recombination is ArF^* and not Ar_2F^*. Recent <u>ab initio</u> calculations by WADT and HAY[11] show that the stable configuration of Ar_2F^* is triangular. So for $Ar_2^+ + F^-$ to form Ar_2F^* the ion trajectory has to be contained in the plane that is normal to the axis of symmetry. Any other trajectory will result in a strong interaction between F^- and Ar^+ and reduce the attractive force between the Ar^+ and Ar resulting in ArF^* formation.

As a result of the experimental results and calculations shown in Fig. 4 one can conclude that the decrease in the fluorescence amplitude with increasing Ar pressure is caused by quenching of ArF^* by Ar. So the ArF^* fluorescence signal S can be written as

$$S = \frac{a N_{Ar}}{1 + (k_{F_2} N_{F_2} + k_{Ar} N_{Ar} + k_{2Ar} N_{Ar}^2)\tau} \tag{1}$$

where α is a constant, τ is the ArF^* radiative lifetime, k_{F_2} is the quenching rate constant of ArF^* by F_2, k_{Ar} and k_{2Ar} are the two and three body quenching rate constants of ArF^* by Ar. N_{F_2} and N_{Ar} are the number densities of F_2 and Ar respectively. Analysis of (1) to obtain the $k\tau$ products has been discussed in detail previously. [7] The curve in Fig. 3 is a plot of (1) using the quenching rate constants obtained by that analysis.

(c) Displacement Reaction

The rate constant for the displacement reaction $Kr + ArF^* \rightarrow KrF^* + Ar$ was obtained by observing the decay of the steady state fluorescence intensity at 1930 $\mathring{A}$ as the partial pressure of Kr was increased. These measurements were made in mixes containing a constant amount of Ar and F_2. The argon partial pressure was 100 torr to minimize the formation of Ar_2^+. In fact, at this low pressure and for an E-beam current of 5 A/cm^2 we have numerically evaluated that Ar^+ is about an order of magnitude greater than Ar_2^+ and the Ar_2^+ channel introduces a variation of about 10% in this comparison.

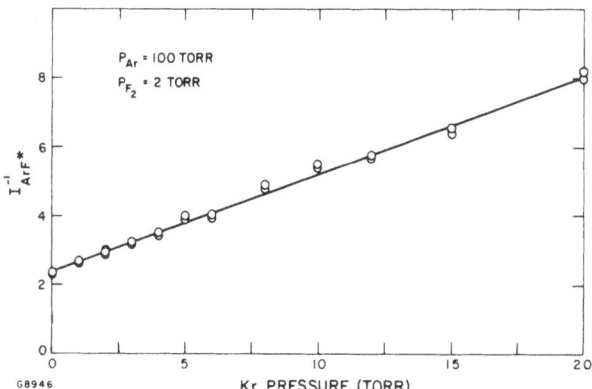

Fig. 5 Stern-Volmer quenching curve for ArF^* with Kr

Figure 5 shows a Stern-Volmer plot of the ArF^* fluorescence data as a function of the Kr partial pressure. From these plots we obtain the half quenching pressure of ArF^* by Kr. The displacement reaction rate constant was also measured by observing the increase in the KrF^* fluorescence amplitude with increasing Kr partial pressure. This measurement gives the same rate constant to within 10% so one can conclude that Kr displaces ArF^* to form KrF^* with a branching ratio near unity.

The radiative lifetime of ArF^* has been calculated to be 4 ns by DUNNING and HAY. [12] Such a short lifetime gives a displacement rate constant

$(Kr + ArF^* \rightarrow KrF^* + Ar)$ of 1.6×10^{-9} cm^3. This rate constant seems unusually large even for this sort of reaction. For example, the kinetically similar alkali-halide reaction $Rb + KF \rightarrow RbF + K$ has a rate constant about an order of magnitude smaller than these rare gas halide displacement reactions. [13] One reason for the difference may be related to the much higher exothermicity of the rare gas halide displacement reactions. [14] Another possible explanation for the rapid displacement rate constant is that at low pressure the Kr atom displaces the Ar atom when the ArF* is in a high vibrational level and thus has a large cross section. At higher pressures the vibrational relaxation of ArF* will proceed more rapidly. Therefore, one might expect the measured displacement rate to decrease with increasing pressure. Preliminary measurements at 200 and 300 torr Ar indicate that this indeed may be the case.

IV. Formation and Quenching of KrF*

(a) Formation of KrF*

In mixtures containing mainly Ar at low total pressures, ArF* is formed first. KrF* is subsequently formed by the displacement reaction as discussed in the previous section. At pressures of about an atmosphere and greater (depending on E-beam current density), molecular argon ion formation becomes important. The Ar_2^+ rapidly charge transfers with Kr to form Kr^+. [15] For lean Kr mixes the Kr^+ recombines with F^- to form KrF^*. As the Kr partial pressure and total mixture pressure are increased, Kr_2^+ will be formed. By experimental measurements similar to those discussed in the previous section, we have shown that Kr_2^+ recombines with F^- to form mainly KrF*. Once KrF* is formed it can radiate or be quenched by the constituents of the gas mixture. The dominant quenching processes and reaction rates are listed in Table 3.

Table 3 Dominant quenching processes of KrF*

Reaction		$k\tau_R(KrF^*)$	$k(\tau_R = 6.5$ nsec$)$	
$KrF^* + F_2$	$\rightarrow$ Products	5×10^{-18} cm^3	7.8×10^{-10} cm^3 sec^{-1}	(1)
$KrF^* + 2Kr$	$\rightarrow Kr_2F^* + Kr$	4.4×10^{-39} cm^6	6.7×10^{-31} cm^6 sec^{-1}	(2)
$KrF^* + Kr$	$\rightarrow$ Products	$\leq 1.1 \times 10^{-20}$ cm^3		(3)
$KrF^* + Kr + Ar$	$\rightarrow Kr_2F^* + Ar$	4.2×10^{-39} cm^6	6.5×10^{-31} cm^6 sec^{-1}	(4)
$KrF^* + 2Ar$	$\rightarrow$ Products	4.6×10^{-40} cm^6	7×10^{-32} cm^6 sec^{-1}	(5)

(b) Quenching of KrF[*]

The rate constant for quenching of KrF[*] by F_2 was measured by observing the KrF[*] fluorescence amplitude versus pressure in binary mixtures of Kr and F_2. The procedure was similar to the measurements of ArF[*] quenching by F_2 described in the previous section. The two body quenching of KrF[*] by Kr and the three body quenching by 2Kr were studied in Kr/F_2 mixes, similar to the analogous case of ArF[*] quenching by Ar as discussed in the previous section.

Figure 6 shows the spontaneous emission spectra in mixtures containing 0.3% F_2, 6% Kr and 93.7% Ar at various total pressures. The uncalibrated spectral intensity scale is approximately logarithmic. At 0.5 atmosphere essentially all of the radiation from the mixture is contained in the KrF[*] $^2\Sigma_{1/2} \rightarrow {}^2\Sigma_{1/2}$ band at 248 nm. However, two other broad bands, containing much less energy, are observable. The first is centered at 415 nm and has been identified with the $^2B_2 \rightarrow A_1$ transition of the excited triatomic Kr_2F^*. [16]

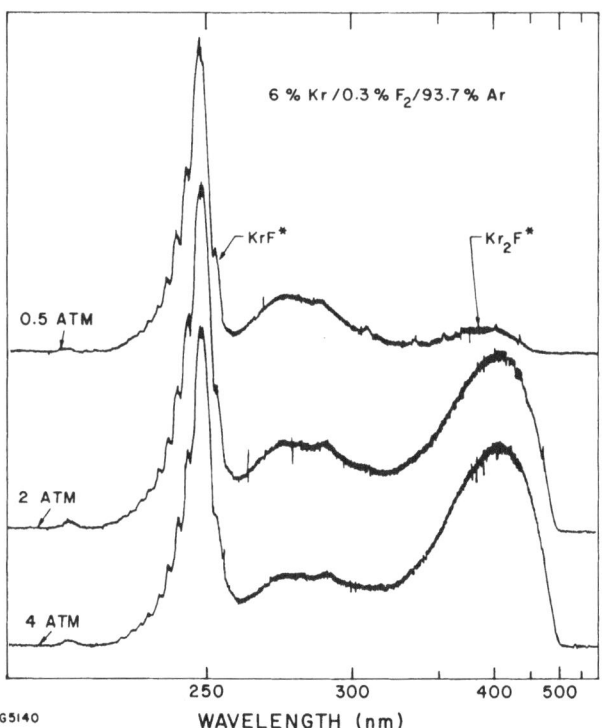

G5140

Fig. 6 Spectra of a 6% Kr mixture at various total pressures

The other broad band, centered roughly at 270-280 nm, is most likely a combination of radiation from the $^2\Sigma\text{-}^2\pi$ band of KrF^* and perhaps radiation from the excited triatomics Ar_2F^* [16] and $ArKrF^*$. Identification of the Kr_2F^* and Ar_2F^* bands was inferred by observing the radiation from binary mixtures of Ar/F_2 and Kr/F_2. From the Kr/F_2 mix we observed the same spectra except that some of the structure in the band centered at 270-280 nm disappeared. From the Ar/F_2 mixture, the spectra showed only a very broad band centered at 290-300 nm which has been identified as Ar_2F^* radiation. The spectrum obtained at a total mixture pressure of 4 atm indicates that, compared with the 0.5 atm spectrum, essentially the same energy is contained in the KrF^* $^2\Sigma_{1/2} \to ^2\Sigma_{1/2}$ band, although the electron beam energy deposited increased by a factor of ~8. This spectrum indicates that most of the additional energy deposited by the E-beam was channeled to Kr_2F^*. This decrease in the KrF^* fluorescence efficiency and increase in the Kr_2F^* fluorescence efficiency could be due to two possible effects:

(1) KrF^* quenching by Ar and Kr, or

(2) Decreasing formation efficiency of KrF^*. Such a decrease is expected if Kr_2^+ recombines with F^- to form KrF^* with a branching ratio <1. Experiments performed with different E-beam currents, similar to those discussed in the previous section, show that Kr_2^+ recombines with F^- to form KrF^* with a branching ratio near unity.

To further substantiate this conclusion, an experiment was performed in which some of the KrF^* formed were deactivated by stimulated transition induced by a KrF^* laser before they could be collisionally quenched. If Kr_2F^* is formed by the quenching of KrF^* as our measurements indicate, then stimulation of the KrF^* transition $(B^2\Sigma_{1/2} \to X^2\Sigma_{1/2})$ by intense radiation at 249 nm should lead to a decrease in the Kr_2F^* fluorescence amplitude. This experiment was performed in a one meter long E-beam pumped laser device. Two quartz windows at the side of the cavity enabled us to look simultaneously at the KrF^* and Kr_2F^* sidelight fluorescence. By blocking and unblocking the end laser mirror, the nonsaturated and saturated sidelight emissions could be compared. Typical results of this experiment are shown in Fig. 7. On the right nonsaturated no lasing sidelight fluorescence pulses of KrF^* and Kr_2F^* are shown. On the upper left same fluorescence pulses when the laser is on are shown. At the bottom left the saturated Kr_2F^* pulse and the laser pulse are shown for timing comparison. From Fig. 7 it is evident that both KrF^* and Kr_2F^* fluorescence intensities are reduced drastically with the onset of the laser. Note that the onset of the saturation for the Kr_2F^* is delayed by about 50 nsec compared to that for KrF^*, as expected. The ratio of KrF^* fluorescence intensity when lasing to the intensity without lasing is for that particular case 0.38 while same ratio for Kr_2F^* is 0.40. Experiments performed with different cavity mirrors and a variety of gas mixtures and pressures gave similar results. The results of this set of experiments confirm that Kr_2F^* is a product of the KrF^* quenching. Therefore, the decrease of the KrF^* fluorescence efficiency with increasing pressure is a result of quenching by Ar and Kr.

LASER ON   LASER OFF

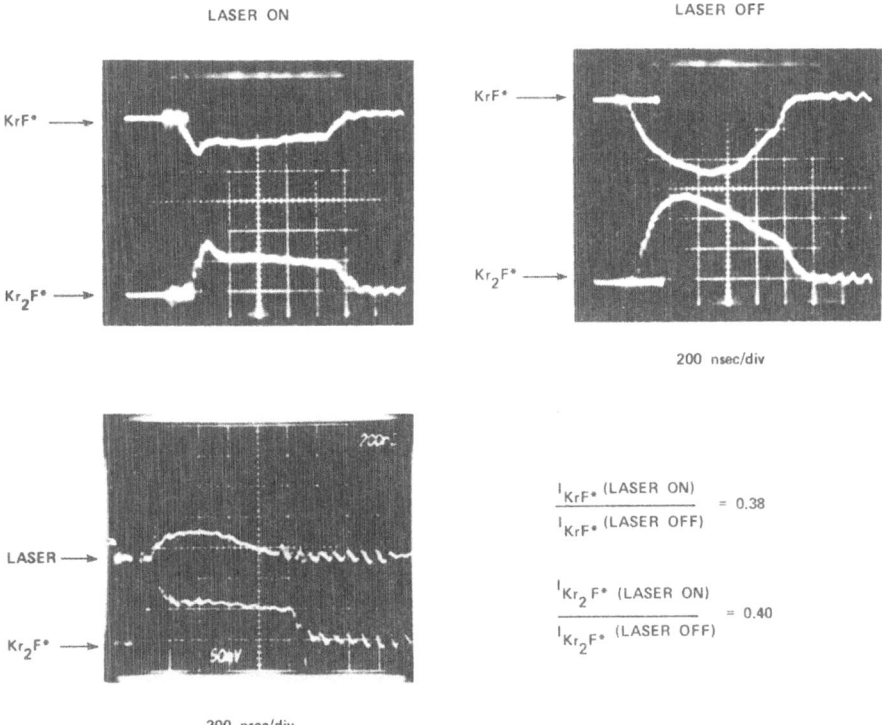

$$\frac{I_{KrF^*} \text{ (LASER ON)}}{I_{KrF^*} \text{ (LASER OFF)}} = 0.38$$

$$\frac{I_{Kr_2F^*} \text{ (LASER ON)}}{I_{Kr_2F^*} \text{ (LASER OFF)}} = 0.40$$

200 nsec/div

<u>Fig. 7</u> Data showing the KrF^* and Kr_2F^* side light fluorescence in the presence and absence of laser flux.

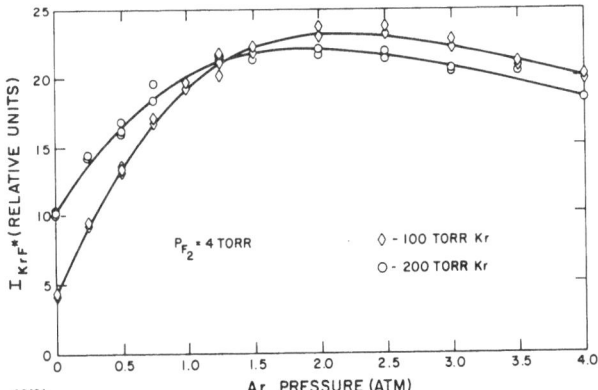

G8484

<u>Fig. 8</u> KrF^* fluorescence signal for $Ar/Kr/F_2$ mixtures containing 4 torr F_2 and 100 and 200 torr Kr. The data was taken varying only the Ar partial pressure. The curves are the calculated fluorescence using the rate constants listed in Table 3

The quenching of KrF^* by processes (4) and (5) in Table 3 were studied in $Ar/Kr/F_2$ mixtures. Reaction (5) was determined by analyzing the dependence of the KrF^* fluorescence on the Ar partial pressure. The F_2 and Kr partial pressures were kept constant for these runs. For such a mix reaction (4) appears as an effective two-body reaction (Kr constant). Because of their different pressure dependence, two and three body processes can be differentiated by an analysis similar to that discussed in detail in [7].

To measure reaction (4) more accurately, the KrF^* fluorescence intensity as a function of the Kr partial pressure was measured. In these measurements the Ar and F_2 partial pressures were kept constant. Fig. 8 shows typical data for the KrF^* fluorescence intensity as a function of Ar partial pressure. The curves represent the predicted pressure dependence of the KrF^* intensity using the rate constants in Table 3.

V. Conclusion

In the preceding sections we have shown that the ArF^* and KrF^* exciplexes can be formed via the ion channel with unit branching. From our measurements we can conclude that the decrease of the fluorescence intensity with increasing pressure is a result of two and three body quenching of the exciplex.

Interception of the precursors (Ar^+, Kr^+, and F^-) of these exciplexes are a negligibly small effect when the experimental conditions are chosen properly. In a laser it is possible to minimize the loss due to the quenching processes by saturating the lasing transition. We have identified the dominant quenching processes of ArF^* ($B^2\Sigma_{1/2}$) and KrF^* ($B^2\Sigma_{1/2}$) and measured their rate constants.

Acknowledgment

We greatly appreciate the expert technical assistance of R. Brochu and J. B. Dodge.

References and Footnotes

1. J. E. Velazco and D. W. Setser, JQE 11, 708-709 (1975).
2. J. J. Ewing and C. A. Brau, Appl. Phys. Lett. 27, 350-352 (1975).
   C. A. Brau and J. J. Ewing, Appl. Phys. Lett. 27, 435-437 (1975).
   S. K. Searles and G. A. Hart, Appl. Phys. Lett. 27, 243-245 (1975).
   E. R. Ault, R. S. Bradford and M. L. Bhaumik, Appl. Phys. Lett. 27, 413-415 (1975).

3. J. A. Mangano and J. H. Jacob, Appl. Phys. Lett. 27, 495-497 (1975).
   R. Burnham, H. W. Harris and N. Djeu, Appl. Phys. Lett. 28, 86-87 (1976).
   C. P. Wang, H. Mirels, D. G. Sutton and S. N. Suchard, Appl. Phys. Lett. 28, 326-328 (1976).
   J. A. Mangano, J. H. Jacob and J. B. Dodge, Appl. Phys. Lett. 29, 426-428 (1976).
4. J. A. Mangano et al (unpublished).
5. R. Hunter (unpublished).
6. J. E. Velazco, J. H. Kolts and D. W. Setser, J. Chem. Phys. 65, 3468-3485 (1976).
7. M. Rokni, J. H. Jacob, J. A. Mangano and R. Brochu, Appl. Phys. Lett. 30, 458-460 (1977).
8. L. R. Peterson and J. E. Allen, Jr. , J. Chem. Phys. 56, 6068-6076 (1972).
9. M. R. Flannery (unpublished).
10. It was determined aposteriori that at 150 torr the Ar quenching of ArF* introduces a 5% error. This error was corrected for subsequently.
11. Willard R. Wadt and P. Jeffrey Hay, Appl. Phys. Lett. 30, 573-575 (1977).
12. Thom. H. Dunning and P. Jeffrey Hay, Appl. Phys. Lett. 28, 649-651 (1976).
13. A. Stolte, A. E. Proctor, and R. B. Bernstein, J. Chem. Phys. 65, 4990 (1976).
14. M. Krauss, N. B. S. , (private communication).
15. D. K. Bohme, N. G. Adams, M. Moselman, D. B. Dunkin and E. E. Ferguson, J. Chem. Phys. 52, 5094 (1970).
16. M. Krauss, N. B. S. , (private communication).
17. Hao-Lin-Chen, R. E. Center, Daniel W. Trainor and W. I. Fyfe, Appl. Phys. Lett. 30, 99 (1977).
18. E. W. McDaniel, V. Cermak, A. Dalgarno, E. E. Ferguson and L. Friedman, (Wiley-Interscience, New York, 1970) pp. 338-339.

Electron Beam Controlled, Neon Stabilized XeF Laser

L.F. Champagne

Laser Physics Branch, Optical Sciences Division, Naval Research Laboratory
Washington, DC 20375, USA

High power and efficiency has been demonstrated for the long pulse electron beam pumped rare gas halide lasers [1-5]. However, in order to scale these laser systems limitations such as foil heating and inefficiencies of electron guns must be overcome. The electron beam controlled XeF laser is less constrained by these limitations and for this reason may be the system of choice in going to high average powers. In this paper we will discuss the operation of the electron beam controlled XeF laser and compare its performance with that of the electron beam pumped laser.

Much of what has been learned about the electron beam pumped XeF laser applies to the electron beam controlled system. The threshold pumping currents required for stimulated emission are significantly lower when neon is used as the diluent in place of argon. This substitution of neon for argon also improves the operation of the laser by reducing the optical absorption in the laser medium. In addition, the electron beam controlled discharge exhibits improved discharge stability when neon is used as the diluent due to its lower ionization efficiency. In an applied electric field energies 4 times that deposited by the e-beam alone increase laser output by a factor of 3.5. However, to date the efficiency for the e-beam controlled system is slightly lower than for the directly pumped e-beam system.

The experimental apparatus, which has been described in more detail previously [1], consists of a 1 meter laser chamber with a 2.2 cm optical aperture. A pulsed rare gas ion laser similar to that described by Simmons and Witte [6] probes the discharge. An argon ion line at 364 nm and a neon ion line at 338 nm are used to measure absorption on either side of line center. On line gain and loss is measured with a discharge pumped XeF laser [7]. The probe pulse is monitored with S-5 photodiodes placed before and after the laser chamber.

Figure 1 is a plot of gain and absorption for the XeF laser as a function of energy deposited into the gas by the electron beam. The energy deposited into the gas is calculated by the method used in Ref. [1]. Absorption is measured in both the pure rare gas and the optimum laser mixture. For the argon diluent laser mixture at maximum output power, the optimum operating pressure is 2.5 atmospheres and the optimum concentration is $Ar:Xe:NF_3::99.5:0.36:0.12$. Since the stopping power of neon is about one-half that of argon, all measurements in neon were performed at 5 atmospheres, in order to keep the energy deposited into the gas by the electron beam equal for the two diluents. The neon diluent composition at 5 atmospheres is $Ne:Xe:NF_3::99.76:0.18:0.06$.

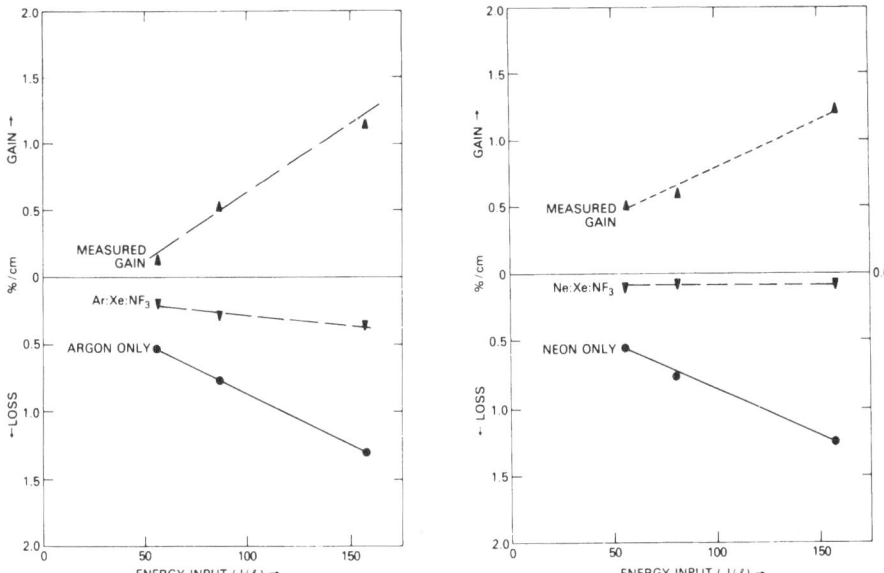

<u>Fig.1</u> Measured gain at 351.1 nm and loss at 338 nm and 364 nm in XeF
laser mixtures as a function of deposited energy for argon diluent
and neon diluent

There is no significant difference in the measured gain at the laser
wavelength for the two diluents. Also, comparable absorption losses are
measured at the wavelength when either pure argon or neon is irradiated
and this absorption is observed to increase with energy input to the
gas. Absorption levels measured on either side of line center are the
same within experimental error for both argon and neon laser mixtures.
The major differences are that the absorption is less in the laser gas
mixture than in the pure rare gases and that the absorption in neon diluent is significantly less than in argon diluent. These data indicate
that a species formed from the rare gas only causes the absorption and
that collisional processes between the rare gas and the laser constituents
either remove an absorbing species or prevent its formation.

Figure 2 plots the optical absorption in both argon and neon as a
function of xenon concentration. Absorption levels are measured at the
laser wavelength. With small additions of xenon to argon and neon a reduction in the optical absorption is observed which is significantly larger
in the neon case. In argon the decrease in absorption is attributed to
charge transfer which reduces the formation of the absorbing species
(presumed to be Ar_2^+) [8]. In neon, Penning ionization provides an additional channel by which a precursor to or an actual absorbing species is
removed. The possible reactions are

$$Ne^* \ (Ne_2^*) + Xe \rightarrow Ne \ (Ne_2) + Xe^+ + e \ . \tag{1}$$

34

For both argon and neon as the concentration of xenon increases the
optical absorption increases indicating that a new absorbing species involving xenon is being introduced. As the concentration is increased
the reaction

$$Xe^+ + Xe + M \rightarrow Xe_2^+ + M,\tag{2}$$

leads to the formation of Xe_2^+ whose cross section for absorption at the
laser wavelength is known to be very large [9]. This strong correlation
between xenon concentration and Xe_2^+ absorption sets an upper limit to
the amount of xenon which can be used in the XeF laser.

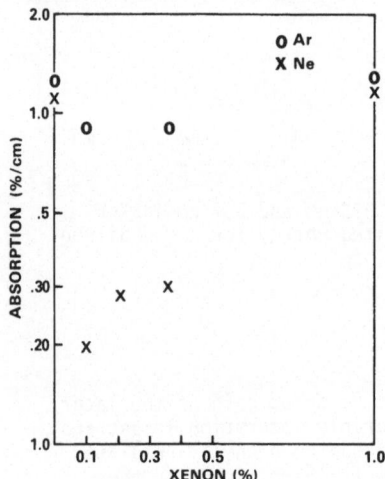

ABSORPTION vs XENON CONCENTRATION

<u>Fig.2</u> Measured absorption at 351.1 nm in both argon and neon as a func-
tion of xenon concentration.

Figure 3 is a plot of output power and efficiency as a function of
applied electric field for the electron beam controlled laser. The operating pressure is 3 atmospheres and data is plotted for one microsecond
optical pulses. In this experiment, the Ne/Xe ratio is kept constant and
the amount of NF_3 is varied.

As the concentration of NF_3 is increased, the maximum value of the
applied electric field for which a one microsecond optical pulse can be
maintained also increases. However, the output power and laser enhancement decreases. At higher applied electric fields, the sustained discharge
cannot be maintained for one microsecond. In the presence of an electric
field the energy extracted is 1.8 J-1^{-1} which is 3.5 times as much energy
as that for e-beam pumping alone. In order to obtain this laser enhancement the energy input to the gas was increased by a factor of five which

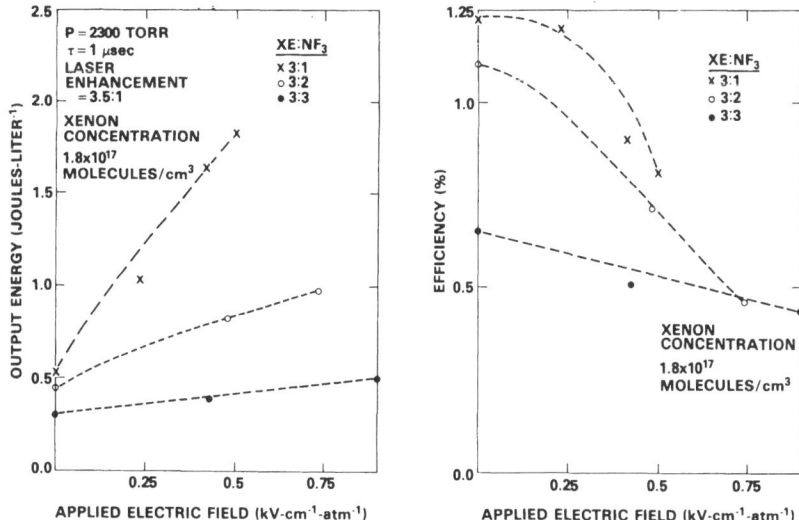

Fig.3 XeF laser output energy and efficiency as a function of applied
       electric field

results in a reduced overall efficiency of the laser system. Most efficient operation and best energy extraction for the e-beam controlled
laser occurs for that concentration which yields the best energy extraction and efficiency for pure e-beam pumping. At 3 atmospheres twice as
much energy can be obtained from the controlled system as from the electron beam pumped laser under optimum conditions.

Table 1 lists the operating characteristics of the electron beam controlled XeF laser when operating at 3 atmospheres. Of note here is that
most of the energy is delivered by the sustainer.

Higher average powers can be obtained at slightly less efficiency
under electron beam controlled operation. Significant laser enhancements
can be achieved from the neon diluent electron beam controlled XeF laser.
When neon is used as the diluent in place of argon threshold pumping
currents are lower and optical absorption at the laser wavelength is significantly reduced.

Table 1   Electron beam controlled XeF laser

Operating Pressure	3.0 ATM
Composition	NF$_3$:Xe:Ne 0.06:0.18:99.76
Optimum Transmission	50%
Output Energy	1.8 J-ℓ^{-1}
Efficiency	0.8%
Measured Gain	1.85%-cm^{-1} [4 ATM]
Laser Enhancement	3.5:1
Energy Deposited by the Gun [E$_g$]	46 J-ℓ^{-1}
Energy Deposited by the Sustainer [E$_s$]	184 J-ℓ^{-1}
$\dfrac{E_g + E_s}{E_g}$	5
Current Density Due to the Electron Beam [Jb]	6 AMPS-cm^{-2}
Current Density Due to the Sustainer [Js]	120 AMPS-cm^{-2}
Js:Jb	20:1

References

1. L. F. Champagne, J. G. Eden, N. W. Harris, N. Djeu and S. K. Searles, *Appl. Phys. Lett.* 30, 160 (1977).
2. R. O. Hunter, C. Houton, J. Oldnettel, postdeadline paper, Third Summer Colloquium on Electronic Transition Lasers, Snowmass, CO, September 1976.
3. L. F. Champagne and N. W. Harris, *Appl. Phys. Lett.* (to be published).
4. L. F. Champagne and N. W. Harris, 1977 IEEE/OSA Conference on Laser Engineering and Applications, Washington, D. C., June 1977.
5. J. H. Jacob, J. A. Mangano, and M. Rokni, 1977 IEEE/OSA Conference on Laser Engineering and Applications, Washington, D. C., June 1977.
6. W. W. Simmons and R. S. Witte, *IEEE J. Quantum Electronics*, QE-6, 648, (October 1970).
7. R. Burnham and N. Djeu, *Appl. Phys. Lett.* 29, 709 (1976).
8. R. O. Hunter, private communication.
9. W. Wadt, private communication.

New Visible Laser Transitions in the Mercury Halides

J.H. Parks

Avco-Everett Research Laboratory, Inc.
Everett, MA 02149, USA

In the last two years visible/ultraviolet laser programs at AERL have led to the discovery and successful development of the rare gas halides. Specifically, 12 J/liter with a 10% intrinsic efficiency has been obtained from KrF^* by pure e-beam pumping and 10 J/liter with a 9.5% intrinsic efficiency from e-beam controlled discharge pumping. Recently another class of visible lasers, the mercury halides, has been discovered. These molecules, like the rare gas halides, have ionic upper levels. Hence, the formation kinetics of the upper laser level should be rapid and efficient. Because the ionization energy of mercury is lower than the rare gases the mercury halides radiate at longer wavelengths than the rare gas halides. Table I lists the four mercury halides and the respective wavelengths of the most intense transitions originating on the upper vibrational level v' and ending on the lower level v''.

Table 1 Mercury halide B-X transitions

Molecule	$\lambda[\mathring{A}]$ (v', v'')	Lased
HgCℓ	5576 (0, 22)	x
HgBr	5018 (0, 22)	x
	4984 (0, 21)	
HgI	4455 (1, 19)	
	4412 (0, 15)	
HgF	7000-9000 approx.	

This paper summarizes the characteristics of the new high power visible lasers operating on the $B^2\Sigma^+_{1/2} \rightarrow X^2\Sigma^+_{1/2}$ transition of HgCℓ [1] at 5576 Å and the same molecular electronic transitions of HgBr [2] at 5018 Å and 4984 Å. In both cases, the upper laser state is ionic in nature and is formed directly by chemical reactions in an e-beam excited mixture of high pressure Ar/Xe and small amounts of Hg, and CCℓ_4 or HBr. The lower laser level is the molecular electronic ground state which is covalent in nature and bound by the order 1 eV. The ionic character of the upper laser level provides the opportunity to utilize the highly efficient formation processes important in lasers such as KrF.

The lasing and spontaneous emission experiments were carried out in an aluminum cell shown in Fig. 1 in which the pressure could be varied up to pressures in the range of 7 atm. The cell temperature was controlled by cartridge heaters positioned to uniformly heat the gas cavity and this temperature was kept below 300°C to allow the use of Viton o-rings. Laser mirrors are housed in heated aluminum holders providing an optical aperature of about 2 cm in diameter and sealed directly to the cell. The cavity mirrors are separated by 24 cm. A Marx generator is used to impulse charge a cold-cathode electron gun to roughly 300 KV for about 150 ns providing a current density of about 100 A/cm^2 in the cell. The e-beam of roughly 1 x 15 cm^2 cross section was injected into the gas transverse to the laser cavity optical axis as shown in Fig. 1. The beam entered the laser cell through a 2-mil cell foil and irradiated a gas volume of about 85 cm^3. The e-beam energy deposition in the gas mixture was deduced from differential pressure measurements.

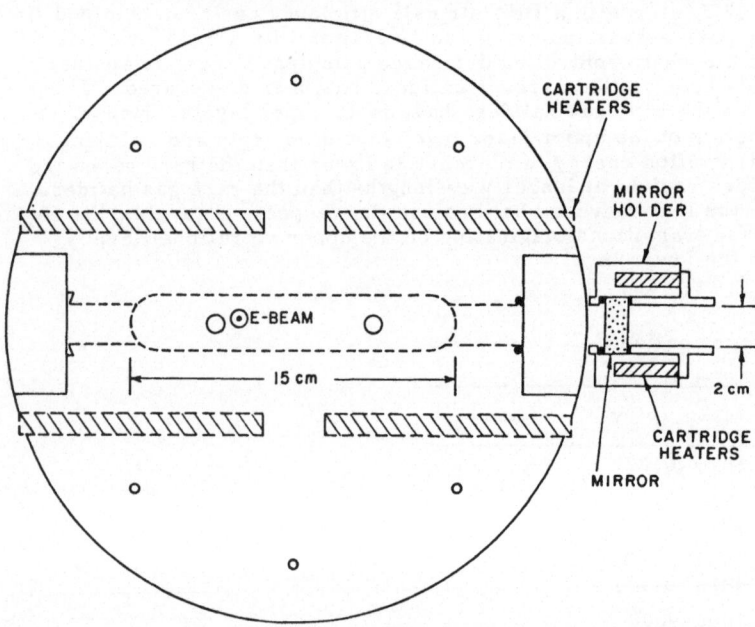

Fig. 1 Aluminum laser cell showing cartridge heater arrangement, cavity and window configuration, and transverse e-beam excitation

Laser action from HgCℓ was obtained in a typical mixture of Ar, Xe, Hg, and CCℓ_4 in the ratio 85.7%/11.1%/2.1%/1.1% respectively at an Ar density of 3 amagats. The cell and Hg reservoir temperatures were 275°C and 260°C respectively. The laser output was viewed at one end with a calibrated planar photodiode (ITT F4000-S5) and at the other end with a 1/2 m Hilger quartz spectrograph. The HgCℓ^* spontaneous emission spectrum, shown

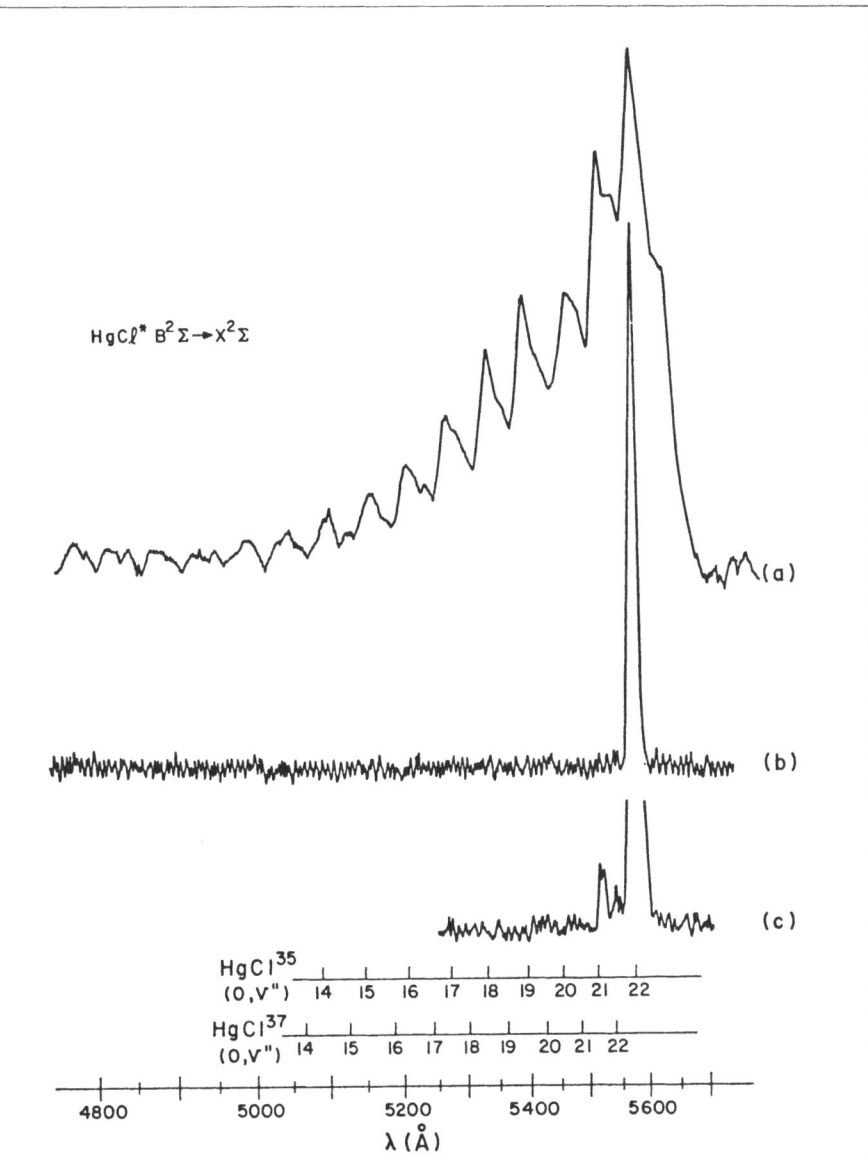

Fig. 2 A comparison of HgCl$\overset{*}{\ell}$ fluorescence and laser spectra.
(a) $B^2\Sigma^+_{1/2} \rightarrow X^2\Sigma^+_{1/2}$ spontaneous emission spectra indicating the
$v'=0$ transitions, (b) the laser spectrum and (c) exposure to several
laser shots to observe weaker lasing transitions

in Fig. 2, was obtained in the presence of these high pressure rare gas mixtures and indicates that the most intense transitions occur between v'=0 and high v'' ground state levels. Based on a previous analysis [3] of the HgCℓ^* spectrum, the laser transition is identified as v'=0 → v''=22 (HgCℓ^{35}) which corresponds to the strongest transition observed in spontaneous emission. The most intense laser output achieved to date is about 1.7 MW, obtained with 50% output coupling. An oscillogram of this data is shown in Fig. 3a. The associated pulse energy of 175 mJ corresponds to an intrinsic laser efficiency of 3.5% considering the e-beam energy deposited in the laser volume was about 4.8 J. Bottlenecking of the v''=22 lower laser level is not limiting the laser pulse width which is roughly 100 nsec. Furthermore, as shown in Fig. 3b, the absence of bottnecking has even been observed using .35% output coupling for which the internal cavity flux was 12 MW/cm^2. This implies a very rapid relaxation of the lower laser level of roughly 1 nsec.

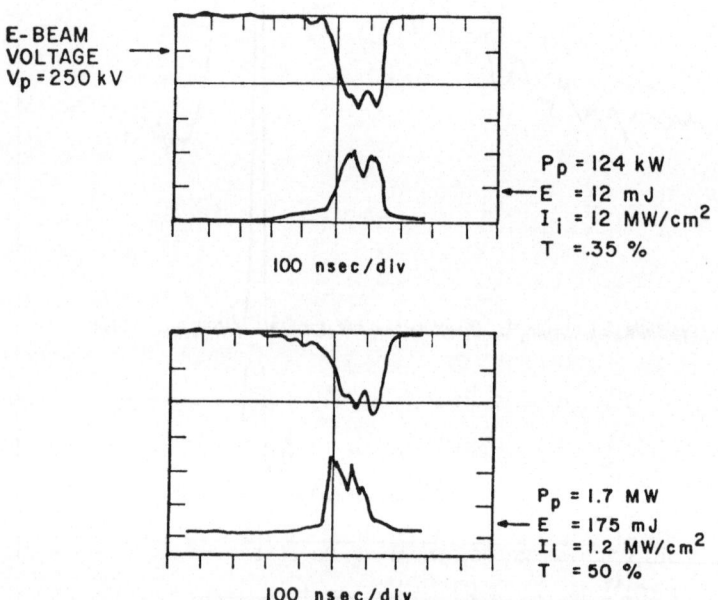

E-BEAM VOLTAGE V_p = 250 kV

100 nsec/div

P_p = 124 kW
E = 12 mJ
I_i = 12 MW/cm^2
T = .35 %

100 nsec/div

P_p = 1.7 MW
E = 175 mJ
I_I = 1.2 MW/cm^2
T = 50 %

Fig. 3 Oscillogram trace showing the e-beam voltage pulse and the corresponding photodiode laser signal for (a) output coupling T = 50%; (b) output coupling T = .35%

Laser action from HgBr was obtained in a typical mixture of Ar, Xe, Hg and HBr in the ratio 86.4%/10.8%/2.0%/.8% respectively at an Ar density of 3 amagats. The laser emission shown in Fig. 4 is tentatively identified [4] as the v'=0 → v''=22 and v'=0 → v''=21 bands at 5018 Å and 4984 Å, respectively, which again corresponds to the strongest transitions observed in spontaneous emission. The output coupling for each cavity mirror was 3%.

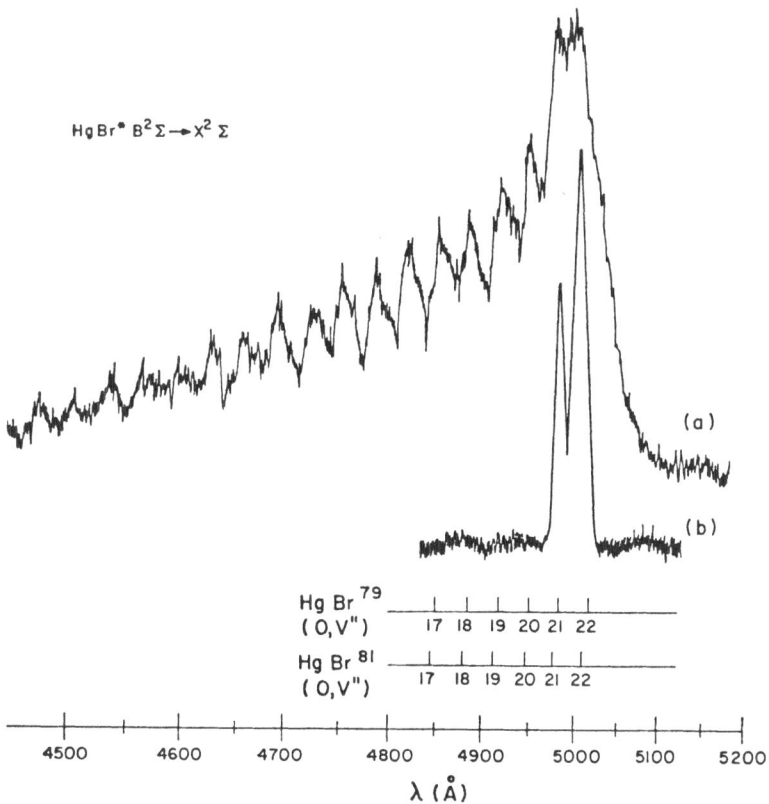

Fig. 4 A comparison of HgBr* fluorescence and laser spectra. (a) $B^2\Sigma^+_{1/2} \rightarrow X^2\Sigma^+_{1/2}$ spontaneous emission spectra indicating the v'=0 transitions, (b) the laser spectrum

The peak laser power of about 50 kW and integrated pulse energy of 3.2 mJ corresponds to a laser efficiency of about .25%. The laser pulse width at half peak intensity is about 60 ns and bottlenecking of the lower laser levels does not appear to be limiting the laser pulse length.

A detailed study of the HgCl and HgBr molecular kinetics with e-beam excitation has not yet been performed. However, the dominant formation channel under these excitation conditions appears to be rapid three-body recombination of Hg$^+$ ions with Cl$^-$ and Br$^-$ ions as summarized in Fig. 5. Similar ion-ion recombination processes can have an effective two-body rate constant of $\sim 10^{-6}$ cm^3/sec at pressures of interest [5]. It is believed that the presence of Xe is necessary for lasing in these gas mixtures because rapid near resonant charge transfer reactions, such as Xe$_2^+$ + Hg $\rightarrow$ Hg$^+$ + 2Xe,

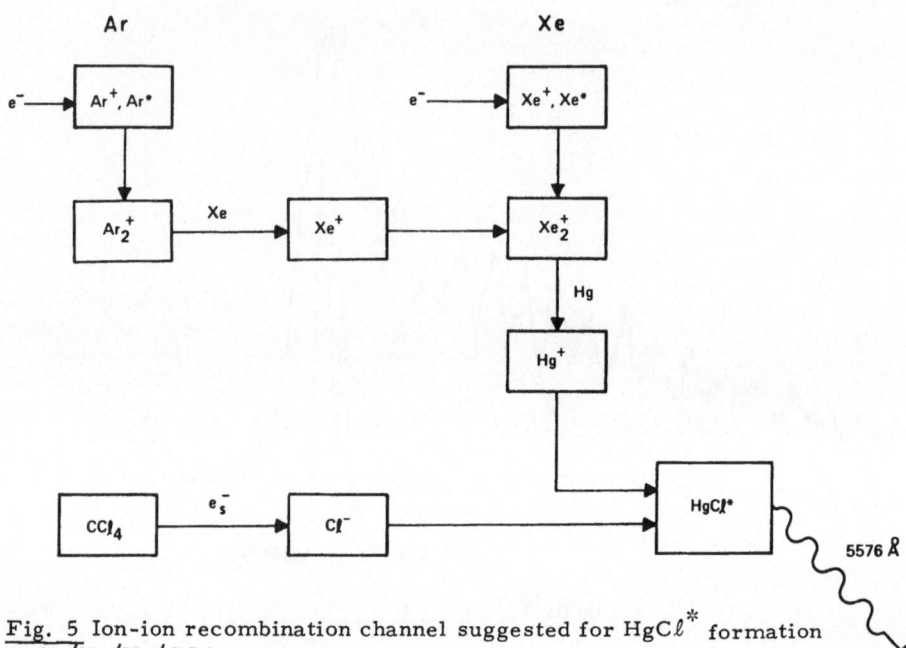

Fig. 5 Ion-ion recombination channel suggested for $HgC\ell^*$ formation in $Ar/Xe/Hg/CC\ell_4$ mixtures

can readily form Hg^+ ions. For example, an analogous reaction involving Ar_2^+ charge transfer to Kr has been measured to have a rate of $7.5 \times 10^{-10} cm^3 /sec$ [6]. The most intense lasing observed in Ar/Xe mixtures is thought to be the result of enhanced Hg^+ formation, possibly via the Ar channel outlined in Fig. 5. The absence of lasing in pure Ar mixtures and the observed weak lasing in pure Xe mixtures suggests the possibility that absorption by Ar_2^* and Xe_2^* may be competing with $HgC\ell^*$ and $HgBr^*$ stimulated emission. The formation kinetics for these absorbing species is outlined in Fig. 6. Note that Ar/Xe mixtures provide a rapid quenching channel for Ar_2^* via Xe collisions having a rate [7] $4.4 \times 10^{-10} cm^3/sec$.

Lasing on the $HgC\ell^*$ and $HgBr^*$ band transitions introduces new high power laser sources at visible wavelengths. The detailed molecular kinetics in these gas mixtures have yet to be established in order to assess the possible efficiency and scalability of these lasers. The optimization of laser mixtures and pumping power presently underway should result in higher values of the intrinsic efficiency. In pure e-beam pumping, the effective quantum efficiency of about 9% is limited by the 26 eV needed to form an argon ion; thus, the 3.8% observed indicates reasonably efficient $HgC\ell^*$ formation. If a discharge were used to pump the Hg^* (6p) metastable levels directly and form HgX^* via neutral reactions the effective quantum efficiency would be roughly 50%. For example, $HgC\ell^*$ formation has been observed in the reaction $Hg^*(^3P_2) + CC\ell_4 \rightarrow HgC\ell^* + CC\ell_4$, and a cross section of $34 Å^2$ was measured [8]. It was originally considered that halogen donor selection would be one of the more critical issues of discharge pumped lasing since the $Hg^*(^3P_0)$ metastable energy (~ 4.7 eV) is close to the average $A-C\ell$ bond strength (~ 3 eV). However, calculations [9] substantiated by measurements [10] in Ar-Hg discharges indicate that electron collisional mixing of the three metastables Hg^* ($^3P_{0,1,2}$) has a large enough cross section ($\sim 19 \ \text{Å}^2$) to maintain the largest population in the highest $Hg^*(^3P_2)$ metastable at 5.43 eV. This considerably eases halogen donor selection for discharge pumping and $CC\ell_4$ is an important candidate. In addition, this provides more flexibility to tailor the choice of halogen donor to the needs of discharge stability.

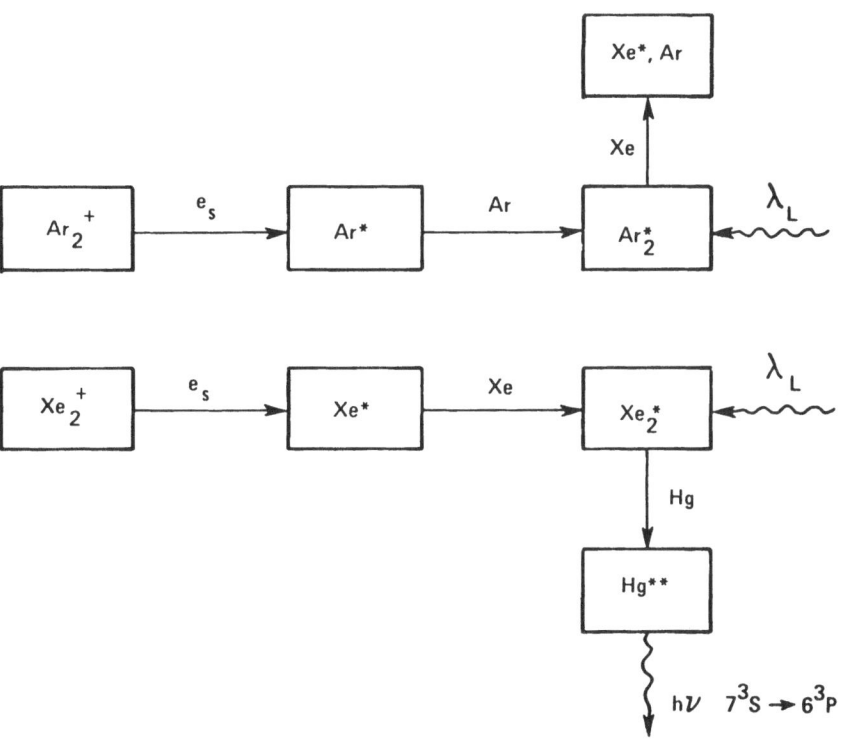

Fig. 6 Suggested formation channels for dominant species which absorb at the laser wavelength λ_L in $Ar/Xe/Hg/CC\ell_4$ mixtures

References

1. J. H. Parks, Appl. Phys. Lett. 31, 192 (1977).

2. J. H. Parks, Appl. Phys. Lett. 31, 300 (1977).

3. K. Wieland, Helv. Phys. Acta. 14, 420 (1941).

4. K. Wieland, Helv. Phys. Acta. 12, 295 (1939); K. Wieland, Z. Elektrochem., 64, 761 (1960).

5. M. R. Flannery, in Case Studies in Atomic Collision Physics 2, edited by E. W. Mc Daniel and M. R. C. Mc Dowell (North-Holland, Amsterdam, 1972), p. 3.

6. D. K. Bohme, N. G. Adams, M. Moselman, D. B. Dunkin, and E. E. Ferguson, J. Chem. Phys. 52, 5094 (1970).

7. R. E. Gleason, T. D. Bonifield, J. W. Keto, and G. K. Walters, J. Chem. Phys. 66, 1589 (1977); A. Gedanken, J. Jortner, B. Raz, and A. Szoki, J. Chem. Phys. 57, 3456 (1972).

8. H. F. Krause, S. G. Johnson, S. Datz, and F. K. Schmidt-Bleek, Chem. Phys. Lett. 31, 577 (1975).

9. B. Yavorsky, Bull. Acad. Sci. U.S.S.R. 9, 233 (1945).

10. C. Kenty, J. Appl. Phys. 21, 1309 (1950).

High Repetition Rate XeF Laser with Gas Recycling

C.P. Christensen

Department of Electrical Engineering, University of Southern California
Los Angeles, CA 90007, USA

The potential of rare gas halide lasers for efficient, high-power generation in the near ultraviolet is well recognized. These devices typically
have been operated at very low pulse repetition rates, however, which has
limited their average power and general utility. In this study we have investigated problems associated with high-repetition-rate operation of a rare
gas halide device and have successfully constructed a discharge-excited XeF
laser capable of operating at pulse repetition rates extending to 200 Hz.

The laser used in the experiments was of a simple Blumlein design as shown
in Fig. 1. The transmission line was constructed of copper-clad fiberglass-
epoxy circuit board material and had a total capacitance of 3 nf. An integral
spark gap pressurized by rapidly flowing compressed air was used in a self-
triggered mode. The device was pulse charged to 13 kV by switching a 10 nf
capacitor across the line with a hydrogen thyratron. Pulse repetition rates
were limited by the high voltage supply to 200 Hz. A gas mixture of $He/Xe/NF_3$
in the ratio 100/3/1 was used to obtain laser output energy of approximately
0.4 mJ over the pressure range 300-760 torr.

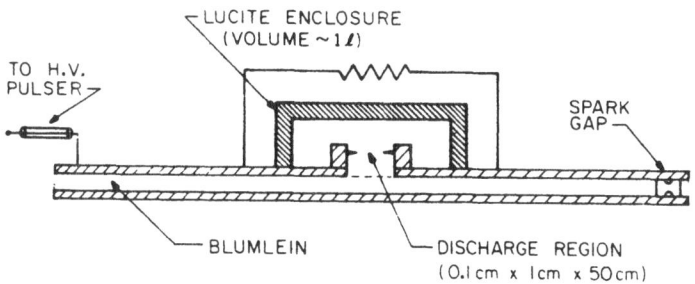

Fig. 1 XeF laser head.

Under static fill conditions energy in the laser pulse was observed to
decrease monotonically as the laser was repetitively discharged. This phenomenon is common to many rare gas halide lasers and apparently is due to

dissociation of NF_3 in the discharge and subsequent wall reactions of the products. Fluorescence from the XeF excimer was monitored in a static gas fill by observing sidelight emission from the active region with a filtered photomultiplier. This fluorescence was observed to decrease exponentially with the number of discharge pulses as seen in Fig. 2. The decay rate was inversely proportional to pressure, but interpretation of this data can only be speculative without additional information.

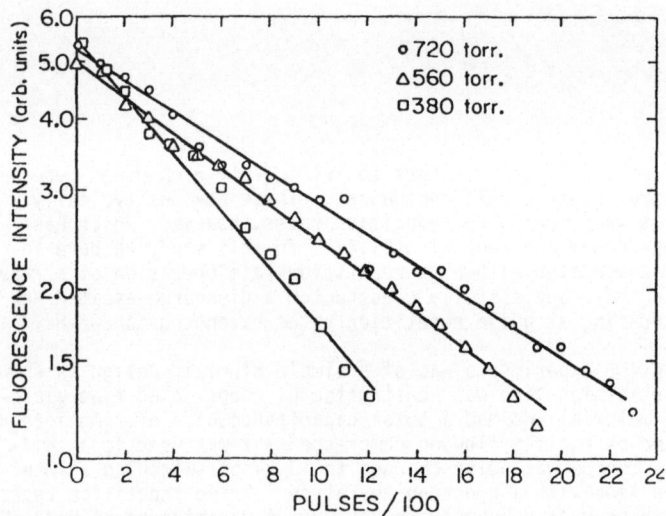

Fig. 2 Decrease of XeF excimer fluorescence in a static gas fill.

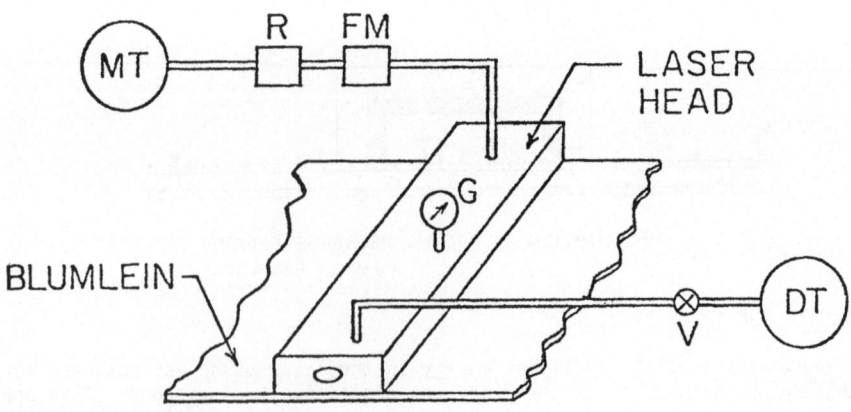

Fig. 3 Longitudinal gas flow arrangement.

To achieve high-repetition-rate operation a longitudinal gas flow arrangement like that shown in Fig. 3 was used. Since the gas mixture is relatively expensive the spent gases were collected in a 100ℓ steel dump tank for later processing. Using this flow apparatus average output power was measured as a function of pulse repetition rate for several gas pressures and flow rates. The results of these measurements, plotted in Fig. 4, could be qualitatively predicted from the static fill data. Best performance was obtained at the highest pressure and flow rate used (720 torr, 6.8ℓ-atm/min). Under these conditions 52 mW of average power was obtained at a pulse repetition rate of 200 Hz.

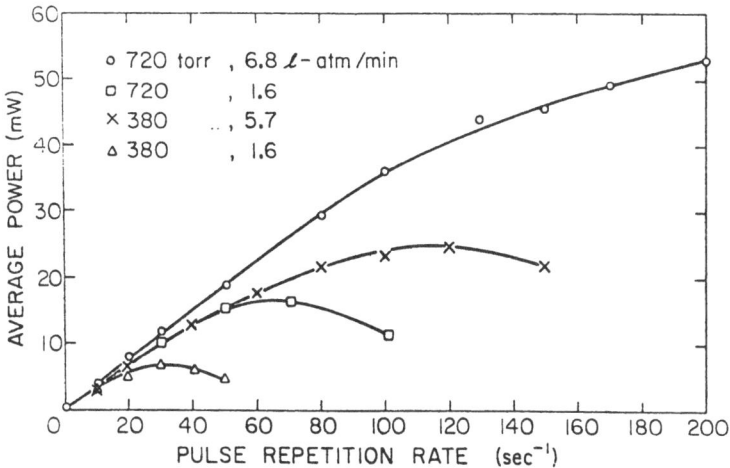

Fig. 4 Average laser output power as a function of pulse repetition rate with a flowing gas mixture.

Reprocessing of the spent gas mixture was undertaken in order to **recover** the costly xenon component. A simple cryogenic processing apparatus, sketched in Fig. 5, was found to be capable of recovering more than 95% of the xenon in a spent mix as well as much of the NF_3. In this process the contaminated gas is passed first through a cold trap at -150°C which removes most of the contaminants and then through a second trap at -196°C which removes xenon and NF_3 from the gas stream. The contents of trap 2 can then be mixed with helium and NF_3 to form a fresh laser gas mixture. With this technique we have been able to recycle the xenon component as many as 10 times with no evidence of cumulative contamination.

Using an IR spectrophotometer we have attempted an analysis of the contents of trap 1 and have identified N_2O, NO_2, and SiF_4 as the primary components. These species suggest the reaction of a nitrogen-fluorine compound such as cis-N_2F_2 with SiO_2 in the fiberglass-epoxy Blumlein dielectric as a possible contamination mechanism [1,2].

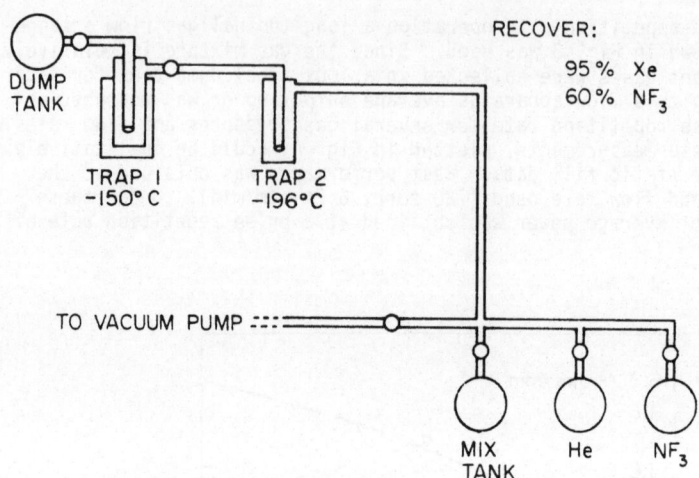

<u>Fig. 5</u> Gas processing station for recovery of Xe and NF$_3$ from a spent gas mix.

References

1. C. B. Colburn, F. A. Johnson, A. Kennedy, K. McCallum, L. C. Metzger, and C. G. Parker, J. Amer. Chem. Soc. <u>81</u>, 6397 (1959).

2. W. Chow, M. Stuke and F. P. Schafer, Appl. Phys. <u>13</u>, 1 (1977).

Simple VUV High Pressure Gas Laser with Coaxial Electron Beam Pumping

G.L. Oomen and W.J. Witteman

Department of Applied Physics, Twente University of Technology
Enschede, The Netherlands

Vacuum ultraviolet stimulated emission of high intensity was obtained in the past by the transverse pumping of high-pressure noble gases by means of high-current relativistic electron beams. The main interest for such laser systems is their potential for obtaining high output powers with high efficiency. In general the lifetime of the upper state of these lasers is in the order of nanoseconds, so that very fast pumping is required. So far it has been reported that short pulses with high current densities were obtained by Blumlein or pulse-line circuits. From a technical point of view such a system is expensive and rather complicated. Therefore one may wonder whether in the case of small systems it is possible to obtain a sufficiently fast discharge from a Marx generator only. This is the main point of the present contribution.

We have constructed a small, convenient VUV laser system with an active medium of about 12 cm length, containing a coaxial diode connected directly to a very fast Marx generator. We have chosen the coaxial excitation scheme because of its very efficient electron pumping [1]. This system allows short-duration low-energy pumping and has produced peak powers in excess of 15 Megawatts in pulses of 90 mJ at 10 atm Xe gas. The experimental

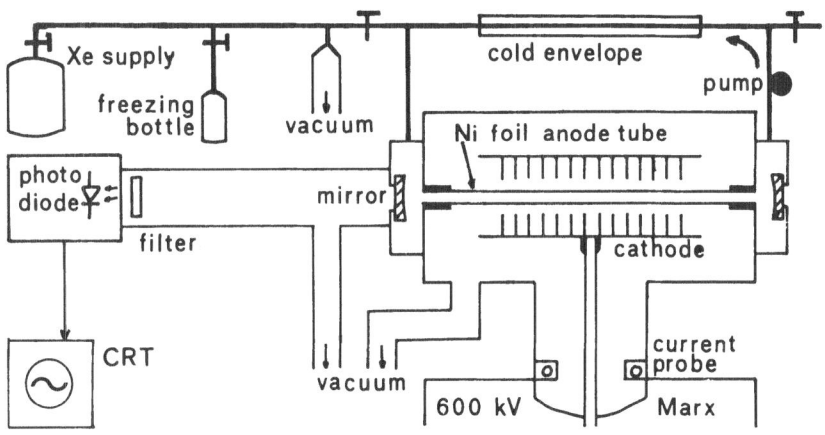

Fig. 1 Experimental set-up

set-up is shown in Fig. 1. The Marx generator, about 1 m height and 30 cm diameter, contains ten stages and has a coaxial design in such a way that each gap is triggered by the preceding firing gap. Only the first gap is triggered externally. The total stored energy of the generator is 120 joules at 600 kVolts. The current was measured with both a Rogowski coil, located between the Marx and the laser, and a PIN diode, placed near the centre of the laser tube. The coil measured the total current from the Marx, whereas the PIN diode measured only the electron current having sufficient energy to penetrate the laser wall and to reach the centre. The latter current can be considered as the excitation current. In Fig. 2 we show the two current

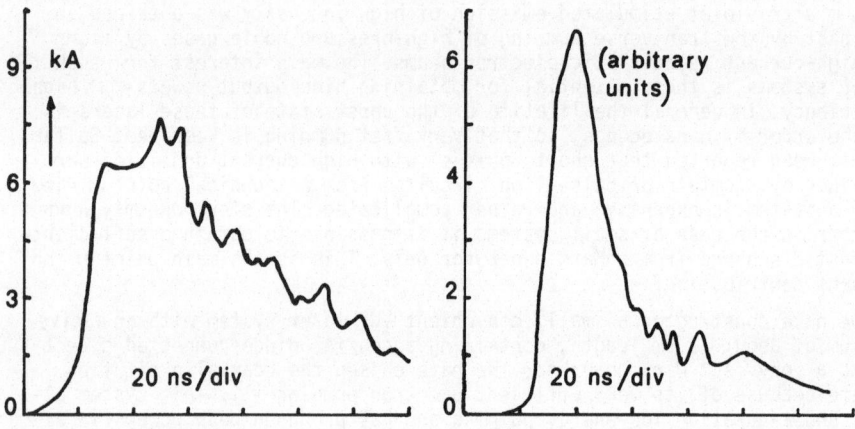

Fig. 2 Current pulses measured with Rogowski coil (left) and PIN diode at the laser axis (right)

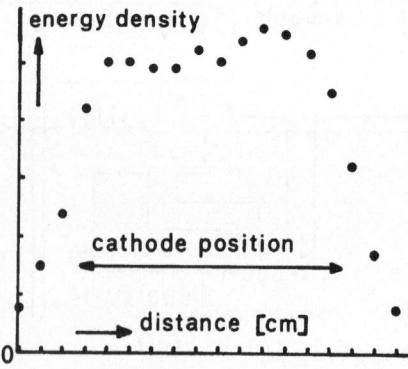

Fig. 3 Electron energy distribution in the tubular anode

measurements. The coaxial field-emission diode consists of a thin-walled nickel tubular anode with a diameter of 6 mm, a length of 20 cm and a wall thickness of 30 microns. The field-emission cathode is an aluminium cylinder of 5 cm diameter with discs of tantalum foil having an inner bore of 25 mm. This provides a more or less uniform excitation of the laser tube over a length of about 12 cm. This was measured calorimetrically. Deviations of the average level were less than 10% and are probably due to small variatons of the anode-cathode distance (Fig. 3). A total energy of 30 J deposited in the tube was measured in this way. From the Berger-Seltzer tables an estimate can be made of the electron energy lost by transversing the tube wall. From this we estimate that about 10 J or 4 J/cm^3 was deposited in the active medium. From this we find that for a total output power of 90 mJ an efficiency of almost 1% is reached.

The spontaneous emission and the stimulated emission where both studied by means of an ITT F4115 photodiode and a Tektronix transient digitizer (Fig. 4). The time constant of the detection system is about 1 nanosecond.

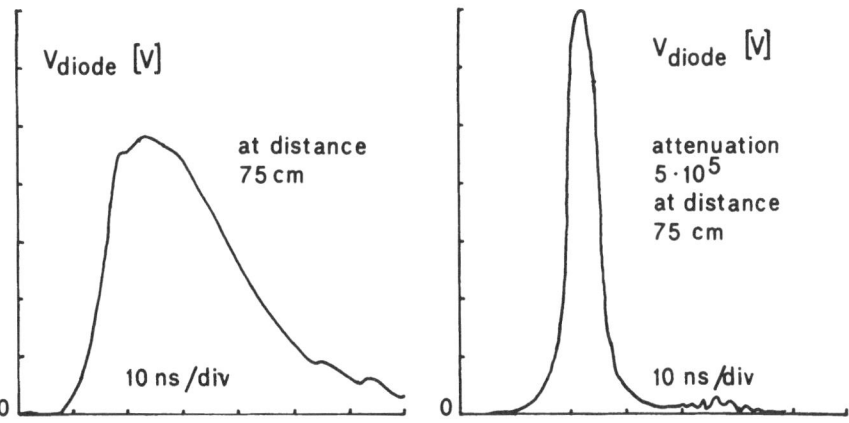

Fig. 4 Spontaneous (left) and stimulated (right) emission from the E-beam pumped Xe gas; time scale 10 ns/div

It is seen that the pulse widths between half maximum intensity points is about 30 and 8 nanoseconds for respectively the spontaneous and the stimulated emission. In order to measure the stimulated emission with the photodiode we used a narrow-band filter at 1700 Å and an air absorption cell for sufficient attenuation.

The laser cavity was formed by dielectric mirrors; one with a reflectivity of 98% and the outcoupling mirror with a reflectivity of 80%. The radius of curvature of both mirrors was 1 meter. The main problem with the mirrors was that for the above-mentioned output power severe burn spots appeared on the end mirror. Therefore we could not generate higher output powers by increasing the input energy. If the charging voltage of the Marx was not higher than 50 kV, the laser output was just below the damage threshold and the output was fairly reproducable; for a typical series starting with 40 mJ the energy slowly diminished to about 25 mJ per pulse after 20 shots. The Xe gas (99.99% purity) could be used for several days by freezing it to liquid nitrogen temperature and pumping it down to 10^{-5} torr before each series of shots. Furthermore, the laser system had a by-pass for circulating the gas through a cold region (a few degrees below 0° C), so that water vapour and other impurities released from the cavity wall could be frozen in.

Although, due to mirror damage, the system could not operate at maximum performance, we believe that this compact and relatively simple device has already proven to be a valuable tool for the study of high-power VUV radiation and that is has the potentiality of obtaining considerably higher output powers.

Reference

1. D.J. Bradley, D.R. Hull, M.H.R. Hutchinson and M.W. McGeoch; Opt. Commun. 11, 335 (1974) and also 14, 1 (1975)

Long Pulse Electron Beam Excited XeF Laser[1]

B. Forestier and B. Fontaine

Institute of Fluid Mechanics, Aix-Marseille University
13003 Marseille, France

Introduction

Recent studies of electron beam excitation of rare gases with molecular gas impurity have shown that this method provides a means for efficient ul- traviolet and visible laser generation[1] , [2] , [3] , [4] . The Laser-Plas- ma Laboratory of Marseille Fluid Mechanics Institute has developed a pro- gram on such high power, high efficiency, electronic lasers with particular focus on long pulse or possible c.w. emission. The type of excitation used is medium current density, long pulse electron beam (E.B.) or E.B. sustained discharge. The gas mixture is either at ambient temperature and medium pres- sure or supersonically cooled to very low temperature and medium pressure. A strong cooling of the mixture may favor a lowering of the quenching and an enhancement of the rate or an increase of the branching ratio for specific reactions useful for a laser effect. Moreover, supersonic flow is also use- ful to remove heat and waste products in the frame of quasi c.w. or high repetition rate pulsed systems.

Characteristics of long duration laser emission of XeF obtained with a short gain length system are described for argon and neon as diluent with emphasis on the laser power enhancement due to use of Ne. Preliminary results concerning XeO_{7S} fluorescence in $Ar/Xe/CO_2/N_2$ mixtures and N_2^+ fluorescence from $He-N_2$ mixtures at very low temperature are also given.

Experimental Set Up

The home made electron gun, of the cold cathode type, is energized by a Marx generator (300 kV-1500 joules) from Physics International (MX31). The electron beam current density through the 1 mm thick titanium window may be varied between 2 A cm^{-2} over 5 μs and 20 A cm^{-2} over 0,5 μs. The window sur- face is 14 x 2 cm^2.

The cell used for experiments at room temperature, without flow, is made of stainless steel and presently allows operation at gas pressures up to 3 atmospheres at full E.B. capacity. A schematic of this laser cell and of the electron gun is shown in Fig. 1.

The cell length is 24 cm, with a 14 cm gain path. The hard coating, multi- dielectric, laser mirrors are integrated with the cell.

The flow cell, made of aluminium with flush mounted electron gun titanium foil, is fed with a supersonic flow generated by a small Ludwieg tube blowing apparatus. This device provides a gas mixture at either 100 or 150 K, depen- ding on the expansion ratio, and a pressure of .3 to .5 atm. The stationary

[1] Work supported in part by French D.R.M.E.

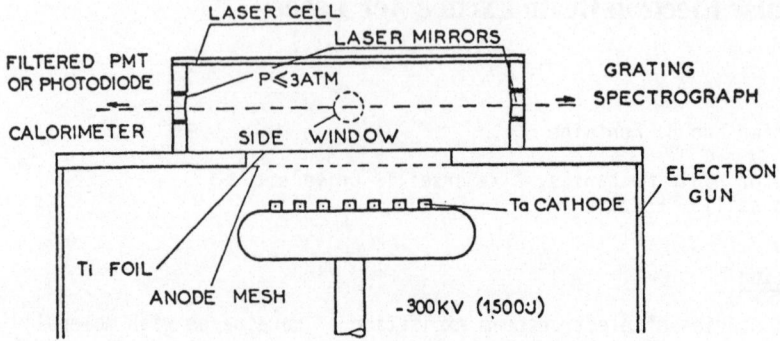

Fig.1 Schematic diagram of the room temperature laser cell

flow duration is about 2 ms. The width of both the cell and the optical cavity is 16 cm and the cell height is 3 cm. A schematic of the flow apparatus is shown in Fig. 2.

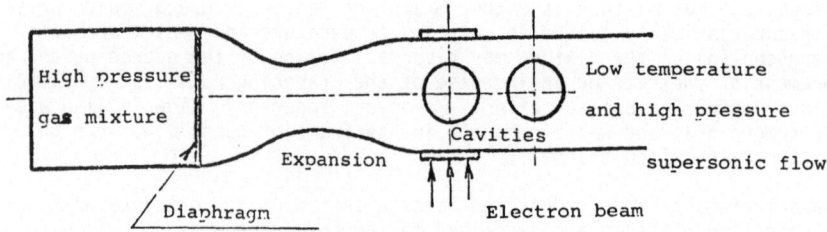

Fig.2 Schematic of the supersonic flow apparatus

For both devices the mixture is prepared in a stainless steel mixing tank outside of the cell. Vacuum pumps allow the maintenance of a pressure of $5 \cdot 10^{-5}$ and $5 \cdot 10^{-4}$ torr respectively in the static and flow cell before filling.

Experimental Results

XeF laser experiments have been performed with the static cell for mixtures of Ar, Xe and NF_3 (1000:4:1) and Ne, Xe and NF_3 (1000:4:1). The pressure was up to 2 atmospheres. The optical cavity used was formed with a 2 m radius mirror of 6 per cent transmittance and a flat mirror of 1 per cent transmittance. The total estimated cavity losses were about 8 per cent for two passes. The optical diagnostics were filtered P.M. or photodiode, calorimeter and grating spectrograph. Typical results of XeF laser experiments are shown in Fig. 3.

We obtained quasi-continuous laser emission on XeF at 3510 and 3530 Å wavelengths for both argon and neon diluent at 1 atmosphere pressure and above. The laser pulse duration was limited by the E.B. current duration only. The threshold conditions for lasing with argon diluent were V = 240 kV and $j_{E.B.}$ = 12 A cm^{-2} for P = 1 atmosphere.

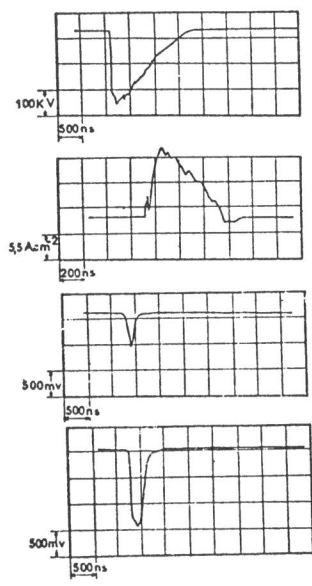

Marx Generator Voltage

$$V_{charge} = 270 \text{ kV}$$

E.Beam Current Density
through 1 mm Ti window

$$V_{charge} = 270 \text{ kV}$$

XeF Laser
Ar:Xe:NF$_3$(1000:4:1)   P=1,1 atm.

$$V_{charge} = 270 \text{ kV}$$

XeF Laser
Ne:Xe:NF$_3$(1000:4:1)   P=1,1 atm.

$$V_{charge} = 270 \text{ kV}$$

Fig.3 Typical temporal profiles from XeF laser experiments

At constant absorbed energy, the threshold for lasing was strongly lowered by use of neon instead of argon as diluent, probably due to the lowering of absorption losses [5] . As an example for V_{gun} = 240 kV, $j_{E.B.}$ = 12 A.cm^{-2} and P=1 atm. of neon we were well above threshold with our short gain length apparatus. Extracted laser power was strongly enhanced by use of the neon diluent. At 2 atm. pressure mixture with neon we obtained 0.4 μs laser pulses of 20 kW peak power. The total laser energy was about 6 mJ for an estimated useful volume of about 10 cm^{-3} which for 2 atm. of argon the extracted energy was three time less. These results are qualitatively similar to that obtained very recently at NRL with a much longer gain length (one meter) [5] . The use of neon as a diluent is very promising for the development of high power, high energy and large volume ($\sim$ 10 liters) laser devices, with Ne/Xe/NF$_3$ mixtures and unstable cavities. A development toward this goal is planned at I.M.F.M. after the fulfilment of the experimental parametric study using the present small device. An other important goal would be to work in the E.B. sustained discharge mode with neon as the diluent in order to further enhance the efficiency of this system. This goal is also planned at I.M.F.M. Very recent successfull results on this mode were presented by L.F. CHAMPAGNE from NRL in a post deadline paper at this colloquium [6] .

Following modelling calculation [7] some fluorescence experiments on the green bands of XeO have been undertaken with our static cell for Ar:Xe:CO$_2$ (100:10:1) mixtures at 1 and 2 atm. The XeO laser is known [8] to give relatively long pulses at very high pressures. We observed that adding large amounts of nitrogen ($\sim$ 5 %) to Ar/Xe/CO$_2$ mixtures produced a twofold increase of peak fluorescence and a tenfold increase of fluorescence energy at 5376 Å (green band of XeO). This effect which has already been observed at SANDIA Lab. [9] for different experimental conditions, seems to be due to the use of the strongly metastable A state of nitrogen as a secondary energy reservoir.

Experiments with the supersonic cooling apparatus are just beginning. We have, however, made some experiments with He-N_2 mixtures and studied fluorescence on the first negative system of N_2 at 4278 Å. Collins [10] has recently shown that this system is favored by low temperature. Experiments made with an He:N_2 (1000:2) mixture at 1 atm. and 120 K showed a slight increase of fluorescence (20 - 30 %) compared to room temperature at the same atom density. These conditions give therefore, from calculations, greater than a threefold gain increase.

The present experiments are only a preliminary phase of a more systematic work on long pulse electronic transition lasers excited by E.B. or E.B. sustained discharges at room temperature or at very low temperature and near atmospheric pressure.

References

1. J.A. Mangano et al., Appl. Phys. Lett. 29, 426 (1976)
2. L.F. Champagne et al., Appl. Phys. Lett. 30, 160 (1977)
3. E.R. Ault et al., Appl. Phys. Lett. 27, 413 (1975)
4. C.H. Fisher and R.E. Center, 3th Summer Colloquium on Electronic Transition Lasers, Snowmass, Colorado, Sept. 1976.
5. L.F. Champagne and N.W. Harris, 5th Conference on Chemical and Molecular Lasers, St. Louis, Miss. April 1977.
6. L.F. Champagne, post deadline paper, 4th Colloquium on Electronic transition Lasers, Munich, June 1977
7. B. Fontaine and B. Forestier, I.M.F.M. report, January 1976
8. H.T. Powell et al., Appl. Phys. Lett. 25, 730 (1974)
9. G.C. Tisone and J.M. Hoffman, Sandia Laboratory Report SAND 74-0425, Albuquerque, N.M., December 1975
10. C.B. Collins et al., Appl. Phys. Lett. 28, 535 (1976)

Part II
Chemical Lasers

Electronic Transition Lasers

S.N. Suchard

U.S. Energy Research and Development Administration
Washington, DC 20545, USA

ABSTRACT

Progress in the development of short wavelength lasers is followed from 1973 to the present. A brief description is given of the various initiation techniques employed as well as of some representative laser systems. This paper is organized in such a way as to provide a historical perspective of research in this area and to provide suggestions for future research.

I. Introduction

Three years ago, at the first of these meetings, there was considerable interest expressed in the demonstration of an efficient electronic transition laser system. Considerable effort had already been successfully put forth in the development of high efficiency infrared lasers, initiated either by an electrical discharge or by a chemical reaction. Consequently, it was felt that the demonstration of an efficient visible laser would require a minimal effort.

At that point in time, literally hundreds of electronic transition lasers already existed [1]; however, for one reason or another none of these lasers could be operated at high electrical efficiencies and still provide volume scalability. An Air Force Panel report [2] reviewing the state-of-the-art in the field of electronic transition lasers was published in 1974. The conclusions reached by this panel and their recommendations were judged sound. Consequently they were used as the framework for the experimental/theoretical investigations that were to follow.

Two basic approaches were utilized for the experimental investigations. In the first approach, the sample gas is irradiated by a high current electron beam which produces large densities of electronically excited states of one or more of the components of the gas. These electronically excited states would then relax to a "relatively metastable" excited state from which they could either lase or undergo a chemical reaction, producing a laser inversion in the product species. The other approach was to investigate chemical reactions of the form,

$$A + BC \xrightarrow{\quad k_1 \quad} AB^* + C \tag{1}$$

$$\xrightarrow{\quad k_2 \quad} AB + C$$

in order to ascertain the possibility of producing an inversion in the AB molecule. This second approach, the operational mechanism of the demonstrated, high power HF chemical laser, is the approach which I will discuss first.

II. Chemically Initiated, Electronic Transition Lasers

Before the first chemical laser was demonstrated [3], a considerable amount of research had already been accomplished in demonstrating that an inversion was produced by the chemical reaction producing the molecule [4, 5]. A compendium of equivalent information available for electronically excited states of diatomic molecules was only completed later [6, 7]. These compendia contain all the available spectroscopic information available at that time; however, at quick glance it becomes evident that the key information, the reactive branching ratios, had not been measured. Since the prime requirement in the demonstration of laser action is the initial production of a laser population inversion, a considerable amount of scientific research was reoriented to the measurement of reactive branching ratios.

Before becoming involved in the actual laser systems themselves, it would be useful to estimate the approximate population inversions that would be required before one would expect laser action to occur in a molecular or atomic system. Using the gain equations derived in Reference 8, for a typical molecular system with a laser cavity length of 1 meter, a population inversion on the order of $10^{12}/cm^3$ is required for 10% gain. For an atomic system, this population inversion requirement is three orders of magnitude less. These required population inversions for laser gain are relatively small, consequently the possibility of obtaining an electronic transition laser pumped by a chemical reaction appears good.

A complimentary analytical study was made in which the diatomic molecular species which would require the minimum population in the upper electronic state for the population of a population inversion were identified [8]. Using the results of this study and the spectroscopic parameters of the excited molecular levels in which there was interest, the minimum reactive branching ratio for system gain could be calculated. Table 1 is a listing of the molecular families which, using the results of this study, appeared to have the greatest chance in producing a chemically-pumped electronic transition laser.

Table I. Potential Candidates for High-Power Chemical Lasers

Chemical family	Examples
Alkaline earth oxides	BeS, BaO
All carbon group molecules	CSe, GeF, SiN
Nitrogen group oxides	NS, PO, AsO
Transition metal halides	ScF, YCl
Molecules predicted to have a large reactive branching ratio by spin correlation rules	CS, BrCl, GeO

During the past three years, a large amount of effort has been devoted to the measurement of reactive branching ratios, radiative lifetimes and quenching rates of electronically excited molecular states [9, 10, 11]. The first chemically pumped electronic transition laser based on the reaction of ground state reactants has yet to be demonstrated; however, reactions of electronically excited atoms have produced electronic transition laser action [12, 13]. This is the topic which we will discuss next.

III. Electrically Initiated, Electronic Transition Lasers

Early studies on the electron excitation of rare gas atoms had indicated that relatively large percentages of the excited atoms passed through the ionized state. Also noted, was that as the sample pressure was increased, the quantity of fluorescence from the rare gas dimers increased markedly. Figure 1 is a schematic representation of the kinetic processes which take place in a sample of high pressure rare gas irradiated by a high current electron beam. Laser emission has been observed from rare gas dimers, where X = Xe, Kr, Ar [7].

KINETIC PROCESSES IN RARE GAS LASERS   FIGURE 1.

PRODUCTION

$$e + X \longrightarrow e + e + X^+$$
$$e + X \longrightarrow e + X^*$$
$$X^+ + X + X \longrightarrow X_2^+ + X$$
$$X_2^+ + e \longrightarrow X + X^*$$
$$X^* + X + X \longrightarrow X_2^* + X$$
$$\boxed{X_2^* \longrightarrow h\nu + X + X}$$

PRODUCTION MECHANISM IS DENSITY DEPENDENT

LOSSES

$$X_2^* + X_2^* \longrightarrow X_2^+ + X + X + e$$
$$h\nu + X_2^* \longrightarrow X_2^+ + e$$
$$e + X_2^* \longrightarrow X_2^+ + e + e$$

LOSS MECHANISM IS EXCITATION DEPENDENT

Fig. 2 presents the potential energy curves for the lowest seven electronic states of the Xe$_2$ dimer. Also indicated in Figure 2 are the various pumping and loss mechanisms inherent to this dimer system. An advantage which these dimer lasers offer is the lack of a bound lower laser level. Consequently, if the upper laser level can be pumped in a time which is short as compared to the spontaneous losses of the excited dimers, then laser threshold can be reached and will be maintained for as long as the pumping continues. The lower laser level, being repulsive, will dissociate in a time comparable to one-half the vibrational frequency of the molecular ($\sim$10^{-13} sec). Therefore, no "bottle-necking" will occur.

FIGURE 2
KINETICS OF Xe$_2$ LASER

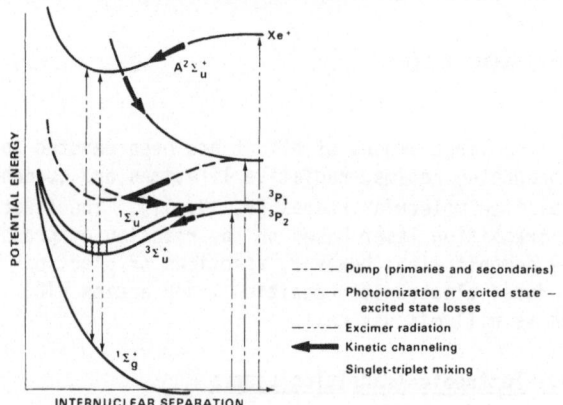

A difficulty with these rare gas excimer lasers which quickly appeared was the high probability of photoionization of the excited dimer (excimer) by the excimer laser radiation. As the intensity of laser radiation was increased in order to saturate the laser transition, the rate of photoionization was increased and, hence, the laser output power was decreased rather than increased. Since the photoionization cross-section is wavelength dependent, and the laser wavelength is fixed by the molecular parameter, the maximum overall electrical efficiency of the rare gas dimer laser was severely limited. The basic technique for the generation of large quantities of electronically excited rare gas atoms and dimers had been developed, now all that was needed was a method by which the wavelength of the laser oscillations could be shifted to a spectral region in which the photoionization of the excited species was negligible.

Two solutions to this dilemma surfaced rather quickly; use the excited dimer radiation as a photodissociation source to produce excited atoms from which laser energy could be extracted or let the excited rare gas atoms chemically react with halogens to produce a population inversion in the product molecule. The first of these solutions results in laser emission from the rare gas oxide molecules [14] and then later in laser emission from the rare gas halide molecules [13, 15].

There are a considerable number of similarities between the rare gas oxide and rare gas halide lasers.

In both laser systems the excited electronic states are bound, or at least stable for a time which is long compared to the vibrational frequency of the molecule. The lower electronic states are unbound, or only very weakly bound.

The differences between these two types of lasers are substantial enough, however, that they fit into completely different uses; rare gas oxides appear ideal as an energy storage medium whereas the rare gas halide laser requires rapid energy extraction for efficient operation. The excited rare gas oxide molecule is formed from a ground state rare gas atom and an electronically excited oxygen atom. The excited rare gas halide molecule, on the other hand, is formed from a ground state halogen atom combined with an electronically excited rare gas atom. Therefore, in the case of the rare gas oxides, the laser emission will occur in the spectral region near the wavelength for the O (^1S-^1D) transition and for the rare gas halides, near the wavelength for the corresponding rare gas atom.

The major difference between these two laser systems in terms of applications, however, is due to the excellent energy storage of properties of the electronically excited oxygen atoms. The $O(^1S$-$^1D)$ transition is highly forbidden. By variations in the rare gas pressure, the stimulated emission cross-section of the rare gas oxide molecule can be varied due to perturbations of the oxygen atom's wavefunction, allowing the laser fusion experimentor to choose the proper cross-section for his particular laser configuration. Large amounts of energy can be stored in the rare gas amplifiers, then rapidly extracted.

Excited rare gas halide molecules have been measured and calculated to have short radiative lifetimes [16]. Consequently efficient energy extraction requires large laser output coupling, with no possibility of energy storage. The overall electrical efficiency for these lasers that has been demonstrated ($\sim$2%) [17] and predicted ($\sim$10%) [18] is sufficiently large to cause a considerable interest for many applications. The laser output wavelengths range from 360nm down to below 200nm. For many applications, the rare gas halide laser wavelengths must be slightly shifted. Research performed at the Los Alamos Scientific Laboratory has shown that stimulated Raman scattering of the rare gas halide laser output by several different molecular gases greatly increases the number of u.v. laser wavelengths available. Figure 3 is a chart of the laser wavelengths available utilizing SRS of the rare gas halide lasers.

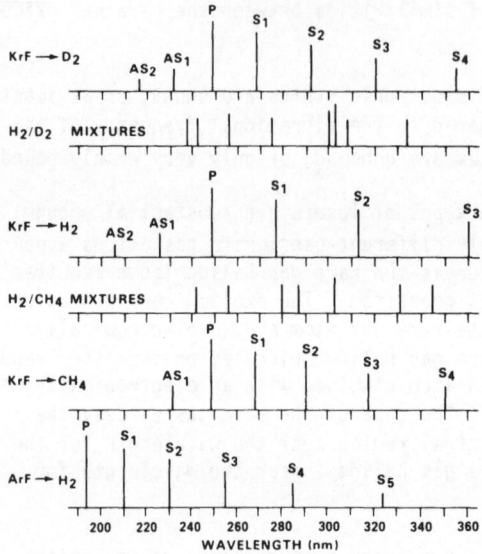

FIGURE 3

**NEW uv WAVELENGTH GENERATED
BY RAMAN SCATTERING**

IV. Conclusions

Efficient visible and ultraviolet lasers now exist. These lasers appear to be volumetrically scalable and operable at high repetition rates. For special applications (e.g., remote operation), a visible laser pumped by a chemical reaction would still be desirable. To accomplish this goal, more basic research into the measurement of reactive branching ratios, quenching rates and spontaneous lifetimes needs to be performed. As specific molecular systems are shown to be attractive, laser experimentation should be initiated.

Extension of electron beam and discharge initiated laser systems has already occurred. Laser action has been demonstrated from the HgCl and HgBr molecules [19]. These molecules are similar to the rare gas halides in that mercury is a psuedo rare gas atom. Following this analogy, future areas of investigation may be centered around other psuedo-rare gas halide molecules (e.g., Cd, Zn-halides).

References

1. Robert J. Pressley, ed., Handbook of Lasers, The Chemical Rubber Co., Cleveland, Ohio, 1971.
2. Leroy C. Wilson, "New Gas Lasers Committee Report on Electronic Transition Chemically and Electrically Excited Lasers," Technical Report AFWL-TR-73-60, Air Force Weapons Laboratory, Kirtland AFB, New Mexico (May 1973).

3. J. V. V. Kasper and G. C. Pimentel, Phys. Rev. Lett. <u>14</u>, 352(1965).

4. G. Karl and J. C. Polanyi, Discussions Faraday Soc. <u>33</u>, 93 (1962).

5. J. C. Polanyi, Appl. Opt., Supplement on Chemical Lasers, 109 (1965).

6. S. N. Suchard, ed., <u>Spectroscopic Data, 1: Heteronuclear Diatomic Molecules, Parts A and B</u> (IFI/Plenum, New York, 1975).

7. S. N. Suchard and J. E. Melzer, eds., <u>Spectroscopic Data, 2: Homonuclear Diatomic Molecules</u> (IFI/Plenum, New York, 1976).

8. D. G. Sutton and S. N. Suchard, Appl. Opt. <u>14</u>, 1898 (1975).

9. G. A. Capelle and J. M. Brown, Jr., J. Chem. Phys. <u>63</u>, 5168 (1975).

10. G. A. Capelle and S. N. Suchard, IEEE, J. Quant. Electron. <u>QE-12</u>, 417 (1976)

11. M. Luria, D. J. Eckstrom, S. A. Edelsten, B. E. Perry and S. W. Benson, J. Chem. Phys. <u>64</u>, 2247 (1976) (and references contained therein).

12. V. S. Zuev, S. B. Kormer, L. D. Mikheev, M. W. Sinitsyn, I. I. Sobelman and G. I. Startsev, JETP Lett. <u>16</u>, 157 (1972).

13. S. K. Searles and G. A. Hart, Appl. Phys. Lett. <u>27</u>, 243 (1975).

14. H. T. Powell, J. R. Murray and C. K. Rhodes, Appl. Phys. Lett. <u>25</u>, 730 (1974).

15. J. J. Ewing and C. A. Brau, Appl. Phys. Lett. <u>27</u>, 350 (1975).

16. T. H. Dunning and P. J. Hay, Appl. Phys. Lett. <u>28</u>, 649 (1976).

17. S. Rockwood, Los Alamos Scientific Laboratory (private communication).

18. J. Daugherty, Avco-Everett Research Laboratory (private communication).

19. J. H. Parks, "Mercury Chloride Laser at 5576Å," presented at the 4th Summer Colloquium on Electronic Transition Lasers, Munich, Germany, June 20-24, 1977 (to be published).

Activation and Deactivation Rates in High-Power Lasers

J.I. Steinfeld

Department of Chemistry, Massachusetts Institute of Technology
Cambridge, MA 02139, USA

1. Introduction

A detailed knowledge of activation and deactivation rates for excited molecular states is essential for the design of molecular lasers and their
development into efficient high-power devices. This is especially true
when one is no longer dealing with the well-known infrared laser systems
discussed by Dr. WILSON at this meeting, but instead is attempting to
devise new systems, based perhaps on molecular electronic transitions in
the visible or ultraviolet portions of the spectrum. Table 1 lists a
few such systems. In this paper, we shall consider some of the mechanisms
for production and relaxation of the energy levels involved in these lasers.

2. Pumping Reactions

The reaction which pumps the HF laser, namely

$$F + H_2 \rightarrow HF(v) + H \tag{1}$$

is well known to generate absolute population inversions between $v = 1$
and 0 and $v = 2$ and 1, and at least partial inversions between $v = 3$ and
2 [15]. The search for reactions producing similar inversions in electronically excited states has, so far, proven to be considerably more
difficult. A typical metal-atom oxidation reaction, such as [16]

$$La(^2D_{3/2}) + O_2\ (^3\Sigma_g{}^-) \rightarrow \begin{cases} O(^3P) + LaO\ (A^2\Pi) \\ O(^3P) + LaO\ (B^2\Sigma^+) \\ O(^3P) + LaO\ (C^2\Pi) \end{cases} \tag{2}$$

has been shown to produce an essentially statistical distribution of electronic states, so that no inversion is possible. In Information-Theoretical terms, the reaction leads to a "zero-surprisal" distribution [17].

Recently, GOLE and co-workers have found several systems which appear
to produce selectively excited electronic states in high yield [18,19].
These are the Group IIIB monohalides, such as ScF produced by oxidation
of Sc atoms by F_2, ClF, or SF_6, and YCl, produced by reaction of Y with
Cl_2 or ClF. An excited $^3\Sigma^+$ state seems to be formed in at least 8-9% yield.
However, neither laser action nor get gain has been demonstrated in such
systems to date. It appears that most reactions of this type preferen-
tially populate high-lying vibrational levels of the electronic ground

Table 1 Electronic Transition Laser Systems

Molecule	Transition	Wavelength [μm]	Method of Excitation	
CN	$A^2\Pi \to X^2\Sigma^+$	1.1 - 1.9	photodissociation	[1]
I_2	$^3\Pi_{2g} \to {}^3\Pi_{2u}$	0.34	e-beam with Ar* transfer	[2,3,4]
	$B^3\Pi_{0u}{}^+ \to X^1\Sigma_g{}^+$	0.6 - 1.5	optical pumping	[5]
N_2	$C^3\Pi_u \to B^3\Pi_g$	0.33	electric-discharge	[6]
	$B^3\Pi_g \to A^3\Sigma_u{}^+$	0.75 - 0.88	electric-discharge	[7]
	$W^3\Delta_u \to B^3\Pi_g$	3.6	electric-discharge	[8]
S_2	$^1\Sigma_g{}^+ \to X^3\Sigma_g{}^-$	1.09	photochemical, $S(^3D)+OCS$	[9]
	$B^3\Sigma_u{}^- \to X^3\Sigma_g{}^-$	0.47 - 0.48	optical pumping	[10]
KrF	$^2\Sigma^+ \to (Kr+F)$	0.25	e-beam, electric discharge	[2,11,12]
ArF	$^2\Sigma^+ \to (Ar+F)$	0.19	e-beam, electric discharge	[13]
HgCl	$B^2\Sigma^+ \to X^2\Sigma^+$	0.56	e-beam	[14]

state of the product molecule; in order to use such reactions to produce
a laser, some way will have to be found to transfer this stored energy
to a suitable acceptor atom or molecule, in such a way that an inversion
between energy levels of the acceptor species can be maintained.

3. Deactivation Rates

In any laser system, deactivation of the upper laser level by collisions
is often the major limitation to available power and efficiency. Our
understanding of these processes has greatly improved during the past
several years; we now have available several models of good predictive
value for both vibrational and electronic relaxation processes.

3.1 Vibrational Deactivation

An extremely useful approach to the problem of vibrational deactivation
rates has been the Information Theoretical analysis of BERNSTEIN and
LEVINE. According to this theory [20,21], vibrationally inelastic
rates can be represented as

$$k(v \rightarrow v';T) = k°(v \rightarrow v';T) \exp(-\lambda_0) \exp(-\lambda_v |E_v - E_{v'}|/kT) \tag{3}$$

where $k°$ is the so-called "prior rate", which can be calculated from a simple phase-space theory, λ_0 is a (temperature dependent) normalization factor which scales the result to a single experimentally determined rate constant, and λ_v is the vibrational "surprisal" parameter. This parameter can be determined by a sum rule on the rates,

$$\frac{\sum_{v'}(E_{v'}-E_v)k°(v \rightarrow v';T) \exp(-\lambda_v|E_{v'}-E_v|/kT)}{\sum_{v'}(E_{v'}-E_{v''})k°(v'' \rightarrow v';T) \exp(-\lambda_v|E_{v'}-E_{v''}|/kT)} = \frac{<<V(eq)>> - E_v}{<<V(eq)>> - E_{v''}}, \tag{4}$$

where $<<V(eq)>>$ is just the equilibrium vibrational energy. An example of such a synthesis is shown in Fig. 1, for the process

$$I_2^*(v = 43) + M \rightarrow I_2^*(v = v') + M. \tag{5}$$

We see that the surprisal-synthesis method, which involves only a brief calculation, represents the experimental data quite as well as an extensive classical-trajectory calculation.

These methods will be particularly useful for generating state-to-state rate constants for V-V deactivation processes such as

$$HF(v_1) + HF(v_2) \rightarrow HF(v_1 \pm m) + HF(v_2 \mp n) \tag{6}$$

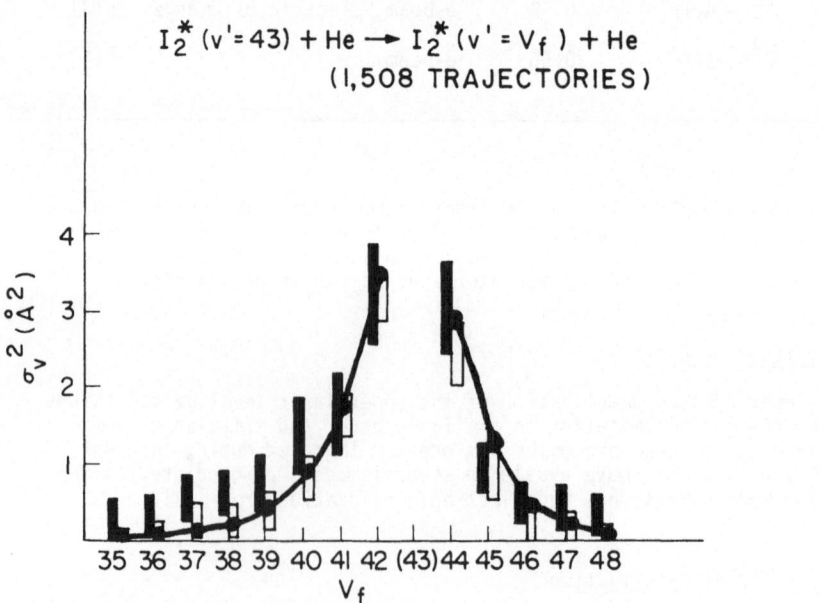

Fig. 1 Cross sections for vibrational energy transfer for monochromatically excited iodine molecules in v = 43. Open bars, experimental results; filled bars, results of classical trajectory calculations [22]; (—●—), Information-Theoretic synthesis with $\lambda_v \simeq 2.5$.

For this system, we have found [23] that, for $[(v_1, v_2) \leq 5]$, there are over sixty channels with significant rates ($k \geq 10^{11}$ cm^3 mole^{-1} sec^{-1}). Essentially, any endothermic process with $\Delta E_v / kT \leq 3$ or any exothermic process with $|\Delta E_v / kT| \lesssim 20$ seems to be possible, with essentially no restriction on the number of quanta exchanged $[(m,n) > 1]$. These results can be validated by comparison to available experimental data, in Table 2.

Table 2   Deactivation Rates for HF(v) by HF(0)

| v | \multicolumn{3}{c}{$k[10^{12}$ cm^3 mole^{-1} sec$^{-1}]$} |
	Experiment [24]	Experiment [25]	Calculated [23][1]
1	1.0	--	1.0
2	12	7.8	13.3
3	11	11.4	12.7
4	16	19.3	13.0
5	35	27.	19.1
6	59	31.	31.
7	--	26.	45.

[1]k is the sum of all V-V and V-(T,R) processes. k(1+0→0+0) chosen for normalization.

These methods will be particularly useful for dealing with systems with a much larger number of levels such as the CO laser ($0 \leq v \leq 30$), or HF with rotational sublevels specified for each vibrational state.

3.2 Electronic Deactivation

It is generally recognized that quenching of excited molecular electronic states proceeds by collision-induced mixing of those states. This may occur as a collision-induced predissociation, when a bound state is crossed by a repulsive state, as in the examples

$$I_2^*(B^3\Pi_{0u}{}^+) + M \rightarrow 2I + M \tag{7}$$

or

$$H_2CO^*(\tilde{A}^1A_2) + M \rightarrow H_2 + CO + M. \tag{8}$$

Quenching may also proceed by a collision-induced perturbation between two bound states, when the rotational manifolds of the two states overlap. Examples are

$$N_2*(B^3\Pi_g) + M \rightarrow N_2*(W^3\Delta_u) + M \tag{9}$$

$$CN*(A^2\Pi) + M \rightarrow CN*(X^2\Sigma) + M. \tag{10}$$

In these latter cases, a limited number of rotational levels act as "doorway states" between the two electronic manifolds.

Several different approaches have been taken to the calculation of probabilities for collision-induced predissociation. A straightforward method involves classical trajectory calculations on the pair of surfaces, with a LANDAU-ZENER crossing probability at the intersection region [26,27]. The problem of multiple crossings makes this method impractical, especially for heavy systems in three dimensions ($I_2* + Xe$). Therefore, an approximate optical model was developed, in which the trajectory takes place on a single surface with crossing represented by an effective absorption probability localized at the intersection region [28]. More recently [29], a semiclassical approximation has been developed, in which the LANDAU-ZENER crossing probability is replaced by analytic continuation of the adiabatic potential energy surfaces. The results of these three models are compared in Table 3.

Table 3   Quenching Calculations for $I_2*(v=43) + M$ by Different Methods

M	1-dimensional Surface Hopping [27]	3-dimensional Optical Model [28]	Semiclassical Decoupling [29]	Experiment
He	0.026	0.025	0.032	0.027
Ar	0.16	0.15	0.20	0.15
Xe	0.27	0.17	0.30	0.45

In the case of collision-induced transfers between two bound states, states of the same J which are nearly resonant in energy are the most strongly coupled. Calculations have been carried out for $N_2(B^3\Pi g)$ relaxing to the $W^3\Delta_u$ and $A^3\Sigma_u{}^+$ states [30], and are in progress for $CN(A^2\Pi)$.

4. Information Theory and Macroscopic Relaxation

In the foregoing discussion, we have seen that Information Theoretical methods can be of great value in systematizing and even predicting inelastic collision rates. An even more important application may be to

the direct computation of the quantities of importance to laser operation, namely, the Populations' Evolution in Time. In this approach
[31], populations of (for example) vibrational levels, $n_v(t)$, are obtained directly without direct use of either energy transfer rates or
of the Master Equation.

In the case of simple vibrational relaxation, these quantities are
given by

$$n_v(t) = \exp[-\beta(t) E_v - \lambda_0(t)]. \tag{11}$$

The time-dependent surprisal parameter $\beta(t)$ is found from a sum rule

$$\frac{\sum\limits_v E_v e^{-\beta(t)E_v}}{\sum\limits_v e^{-\beta(t)E_v}} = <V(\infty)> + [<V(t_0)> - <V(\infty)>]e^{-(t-t_0)/\tau} \tag{12}$$

where $<V(\infty)> = <<V(eq)>>$, the equilibrium average vibrational energy,
$<V(t_0)>$ is the initial non-equilibrium vibrational energy, and τ is the
bulk vibrational relaxation time. The other parameter is determined
from a simple normalization condition,

$$\lambda_0(t) = \ln [\sum\limits_v e^{-\beta(t)E_v}]. \tag{13}$$

Calculations using this method have been carried out for vibrational
relaxation in HF [23,32] and in CO [17]. In both cases, the results
are indistinguishable from a Master-Equation solution using the best
available vibrational deactivation rates, or from the experimental
time-dependent distributions. This method, if applied to the kinetic
modelling codes currently in use for chemical lasers, holds the promise of an enormous saving in the computer time required to run these
codes. Indeed, analyses which are not now feasible, such as inclusion
of rotational disequilibrium, could become possible through the use
of Information-Theoretical modelling techniques.

Acknowledgements

This paper reflects work supported by the U.S. Air Force Office of Scientific Research, Grant AFOSR-75-2758.

References

1. G.A. West and M.J. Berry: J. Chem. Phys. 61, 4700-4716 (1974)

2. R.S. Bradford, Jr., and E.R. Ault, M.L. Bhaumik: Electronic Transition Lasers: Proceedings of the 2nd Summer Colloquium (J.I. Steinfeld,
   ed.)(MIT Press, Cambridge, Mass., 1976) pp. 211-213

3. A.K. Hays, J.M. Hoffman, G.C. Tisone: J.I. Steinfeld, ed., op. cit.,
   pp. 214-216

4. J.J. Ewing and C.A. Brau: J.I. Steinfeld, ed., op. cit., pp. 221-
   224

5. J.B. Koffend and R.W. Field: J. Appl. Phys. (in press)

6. T. Kasuya and D.R. Lide, Jr.: Appl. Optics $\underline{6}$, 66-80 (1967)

7. L.E.S. Mathias and J.T. Parker: Appl. Phys. Letts. $\underline{3}$, 16-18 (1963)

8. R.A. McFarlane: IEEE J. Quantum Electronics $\underline{QE-2}$, 229-232 (1966)

9. V.S. Zuev, L.D. Mikheev, V.I. Yalovoi: Soviet Quantum Electronics $\underline{2}$, 799-806 (1975)

10. S.R. Leone and K.J. Kosnik: Electronic Transition Lasers II: Proceedings of the 3$\underline{rd}$ Summer Colloquium (L.E. Wilson, S.N. Suchard, and J.I. Steinfeld, eds.) (MIT Press, Cambridge, Mass., 1977) pp. 298-301

11. G.C. Tisone, A.K. Hays, J.M. Hoffman: J.I. Steinfeld, ed., op. cit., pp. 191-194

12. C.A. Brau and J.J. Ewing: J.I. Steinfeld, ed., op. cit., pp. 195-198

13. T.R. Loree, P.B. Scott, R.C. Sze: L.E. Wilson et al., eds., op. cit., pp. 35-45

14. J.H. Parks: these Proceedings

15. N. Cohen and J.F. Bott: A Review of Rate Coefficients in the H_2-F_2 Chemical Laser System, Report SAMSO-TR-76-82, Aerospace Corp., Los Angeles, Calif. (April, 1976)

16. J.L. Gole and C.L. Chalek: J. Chem. Phys. $\underline{65}$, 4384-4395 (1976)

17. R.D. Levine and A. Ben-Shaul: Thermodynamics of Molecular Disequilibrium, to be published in Chemical and Biochemical Applications of Lasers, Vol. 3 (C.B. Moore, ed., Academic Press, New York)

18. J.L. Gole and C.L. Chalek: J. Chem. Phys. (to be published)

19. J.L. Gole: L.E. Wilson et al., eds., op. cit., pp. 136-165

20. M. Rubinson and J.I. Steinfeld: Chem. Phys. $\underline{4}$, 467-475 (1974)

21. I. Procaccia and R.D. Levine: J. Chem. Phys. $\underline{63}$. 4261-4279 (1975)

22. M. Rubinson, B. Garetz, J.I. Steinfeld: J. Chem. Phys. $\underline{60}$, 3082-3097 (1974)

23. C. Clendening, J.I. Steinfeld, L.E. Wilson: Information Theory Analysis of Deactivation Rates in Chemical Lasers, Report AFWL-TR-76-144 (October, 1976)

24. M. Kwok and R. Wilkins: private communication

25. P.R. Poole and I.W.M. Smith: to be published

26. R.K. Preston and J.C. Tully: J. Chem. Phys. 54, 4297-4304 (1971); J. Chem. Phys. 55, 562-572 (1971)

27. B. Garetz, M. Rubinson, J.I. Steinfeld: Chem. Phys. Letts. 28, 120-124 (1974)

28. B.A. Garetz, L.L. Poulsen, J.I. Steinfeld: Chem. Phys. 9, 385-391 (1975)

29. G.L. Bendazzoli, M. Raimondi, B.A. Garetz, T.F. George, K. Morokuma: Theoretica Chim. Acta (in press)

30. B.A. Garetz, J.I. Steinfeld, L.L. Poulsen: Chem. Phys. Letts. 38, 365-369 (1976)

31. I. Procaccia, Y. Shimoni, R.D. Levine: J. Chem. Phys. 65, 3284-3301 (1976)

32. I. Procaccia and R.D. Levine: J. Chem. Phys. 62, 3819-3820 (1975)

CW Laser Emission at 3.8μm and 10.6μm Based Upon the Chemical Generation of Chlorine Atoms

S.J. Arnold, K.D. Foster, D.R. Snelling, and R.D. Suart

Defence Research Establishment Valcartier
Quebec, Canada

1. Introduction

DF and HCl lasers which emit in the 3.6 - 4μm region are well suited to laser applications which require high atmospheric transmission. Laser emission at 10.6μm may also be obtained by transfer from DF_v or HCl_v to CO_2. The DF laser system has received the most attention recently because of the greater ease in producing large flows of vibrationally excited DF. The straight chain reaction between D_2 and F_2 proceeds rapidly at room temperature through the "cold" reaction, $F + D_2 \rightarrow DF(v<4) + D$, followed by the "hot" reaction $D + F_2 \rightarrow DF(v\leq14) + F$. The analogous reaction between H_2 and Cl_2 is not sufficiently rapid at room temperature. Even at elevated temperatures the cold reaction, $Cl + H_2 \rightarrow HCl + H$, is nearly thermoneutral and produces mostly $HCl(v=0)$. Consequently, another method of producing vibrationally excited HCl is desirable.

The reaction of NO with ClO_2 was used in this work as a purely chemical source of chlorine atoms with the fast reaction $Cl + HI \rightarrow HCl(v\leq4) + I$ as the laser pumping reaction.

2. Reaction Kinetics

The use of the reaction of NO with ClO_2 as a Cl atom source in chemical lasers was recently reported by our laboratory [1]. A three step branch-chain mechanism is involved:

$$NO + ClO_2 \rightarrow NO_2 + ClO \qquad k_1 = 3.4\times10^{-13}\ cm^3s^{-1} \quad [2] \qquad (1)$$

$$NO + ClO \rightarrow NO_2 + Cl \qquad k_2 = 1.7\times10^{-11}\ cm^3s^{-1} \quad [2] \qquad (2)$$

$$Cl + ClO_2 \rightarrow 2\ ClO \qquad k_3 = 5.9\times10^{-11}\ cm^3s^{-1} \quad [2] \qquad (3)$$

This reaction sequence leads to the production of ClO if $[NO]_0 = [ClO_2]_0$ or to Cl atoms if $[NO]_0 = 2[ClO_2]_0$. This kinetic scheme is referred to as the pre-pumping chemistry. Because of the chain branching step, (3), the overall rate of reaction may far exceed that expected on the basis of reactions (1) and (2) alone. In fact, the time for 90% completion of the reaction is given by

$$\tau_{90\%} = \frac{7.9\times10^{11}}{[ClO_2]_0} \text{ (seconds)} \qquad (4)$$

where $[ClO_2]_0$ is the initial concentration of ClO_2 in cm^{-3}. This characteristic time refers to the time required for CℓO radical production, if $[NO]_0 = [CℓO_2]_0$ or to the time required for Cℓ atom formation if $[NO]_0 = 2[CℓO_2]_0$.

For effective use of the NO/CℓO$_2$ reaction scheme under normal laser operating conditions of a few torr pressure and room temperature, the chain carriers Cℓ and CℓO must not be depleted by loss processes. The most important loss process for both Cℓ and CℓO is the termolecular reaction with NO$_2$, an unavoidable stable product of the pre-pumping reactions. For Cℓ atoms, the NO$_2$ acts catalytically via:

$$Cℓ + NO_2 + M \rightarrow CℓNO_2 + M \tag{5}$$

$$CℓNO_2 + Cℓ \rightarrow Cℓ_2 + NO_2 \tag{6}$$

Because reaction (6) is fast, the Cℓ atoms disappear with an effective termolecular rate coefficient of 1.4×10^{-30} cm^{6}s^{-1} [2]. Termolecular encounters between NO$_2$ and CℓO radicals, however, probably result in a simple conbination:

$$CℓO + NO_2 + M \rightarrow CℓNO_3 + M. \tag{7}$$

The rate coefficient for CℓO loss in the presence of NO$_2$ was measured in our system to be 10^{-31} cm^{6}s^{-1}, i.e. 14 times slower than the analogous loss for Cℓ atoms.

A consideration of the kinetics of the NO/CℓO$_2$ system therefore leads to the preliminary conclusions that a) at typical laser operating pressures of a few torr, production of substantial flows of Cℓ and/or CℓO is possible and b) it should be possible to produce non-equilibrium flows of ClO radicals at pressures substantially higher than that possible for Cl atoms because of reaction (7) is much slower than the combnation of reactions (5) and (6).

3. Flow Laser Considerations

From the above kinetics it appears that a practical laser exploiting the NO/CℓO$_2$ reaction coupled with the Cℓ + HI pumping reaction can be operated in at least two kinetic regimes. In fact, the pre-pumping and pumping chemistry may be accomplished in three ways. These were referred to previously as chemical modes [1] and are summarized in the simple flow laser schematic below.

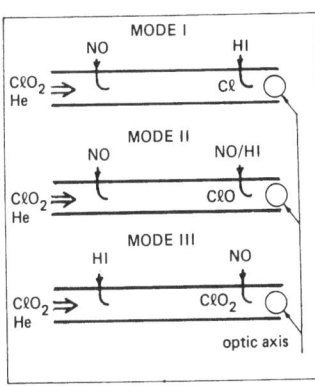

Fig. 1 Gas injection for different chemical modes of a NO/CℓO$_2$/HI flow laser

The modes are distinguished by the form of the chlorine containing species at the point of HI addition. In Mode I, $C\ell$ atoms are produced after the addition of NO through the injector #1 and are subsequently reacted with HI at injector #2. In Mode II, only enough NO is added at injector #1 to form $C\ell O$ radicals which are subsequently reacted in a non-branching reaction at injector #2, first with NO to form $C\ell$ atoms, then with HI to form $HC\ell(v\leq4)$. In Mode III, all the NO and HI are added together at injector #2. Mode I, involving $C\ell$ atom production, is the simplest concept for low pressure operation whereas Modes II and III are preferrable at higher pressures where loss of $C\ell$ atoms would be excessively rapid. The possibility of operation in Mode III is not at all obvious because here the pumping reaction of $C\ell$ + HI competes for $C\ell$ atoms with the chain branching step, $C\ell + C\ell O_2 \rightarrow 2\ C\ell O$. Therefore, $C\ell O$ is formed strictly by the rather slow reaction (1) without much chain branching and thus requires significant excess NO to achieve a sufficiently rapid reaction rate.

The pre-pumping reaction sequence is unaffected by the presence of CO_2 when the laser is operated as an $HC\ell/CO_2$ transfer laser at 10.6μm.

4. Laser Experiments at 3.8μm and 10.6μm

Continuous wave laser experiments employing transverse flow were performed on two devices similar to that described in reference 3. The smaller device (MKI) consisted of a rectangular flow channel 1.5 cm x 14 cm. The total pressure was 2.4 torr and the average linear flow velocity 160 ms^{-1}. The larger device (MK2) had a rectangular flow channel 1 cm x 40 cm. The total pressure was 2.4 torr and the velocity 220 ms^{-1}. The reagent gases were injected into the main flow stream through needle injector arrays [3]. For the MK2 device, mixing was improved by adding extra holes to the injector needles near the centerline of the main flow channel to compensate for the higher centerline gas velocity. Both devices employed $C\ell O_2$ flows of approximately 40 SCCM per sq. cm. of total channel cross sectional area.

The laser cavity was composed of a stable resonator with 4% (MKI) and 10% (MK2) output coupling. With this configuration and with Mode I operation, the MKI device gave maximum multiline output power of 4 W and the MK2 device 13 W.

The optimum performance characteristics of the MK2 transverse flow laser are given below.

Table 1   MK2 Laser Performance

	$HC\ell$	$HC\ell/CO_2$
helium [SCCM]	103600	117600
$C\ell O_2$ [SCCM]	1500	1700
NO #1 [SCCM]	4500	1120
NO #2 [SCCM]		9700
HI [SCCM]	1950	2000
CO_2 [with NO #1]		20800
total pressure [torr]	2.4	4 - 6
wavelength [μm]	3.6 - 4	10.6
total power [W]	13	5
chemical efficiency [%] (based on total ΔH pumping)	8	4

For the MKI device, laser emission was observed for eight P branch lines of the HCℓ^{35} isotopic species. About 60% of the laser power was from the v=2→1 vibrational band. Preliminary measurements of the spectral output of the MK2 device showed HCℓ laser emission on 17 lines. Although not resolved, lines from HCℓ^{37} are expected to lase because of the higher gain of the MK2 device.

Successful HCℓ laser action was also observed for chemical Mode II and III. For experiments performed on the MKI device, the maximum multiline HCℓ laser power was in the approximate ratio 1:0.7:0.5 for chemical Modes I:II:III respectively. The lower power realized from Modes II and III is a consequence of the relatively low laser operating pressure (2.4 torr) where Cℓ atom loss is small. This situation is expected to reverse when future higher pressure experiments are performed.

5. Summary and Conclusions

A transverse flow laser operating at either 3.8μm or 10.6μm was designed using the reaction of NO with CℓO$_2$ as a purely chemical chlorine atom source and the reaction of Cℓ with HI as the laser pumping reaction. The HCℓ power observed of 13 W represents a laser efficiency of 8% (based on total rection exothermicity). The power is equivalent to 366 Jg^{-1} of atomic chlorine. This, to our knowledge, is the most efficient purely chemical laser so far reported.

Laser operation in three distinct kinetic modes was demonstrated. Two of these modes have low relative Cℓ atom concentrations and show promise of scaling to higher pressures.

The HCℓ and HCℓ/CO$_2$ lasers should be considered along with the DF and DF/CO$_2$ lasers for cw laser applications requiring efficient atmospheric transmission.

6. References

1 S.J. Arnold, K.D. Foster, D.R. Snelling and R.D. Suart, Appl. Phys. Let. 30 657-639 (1977).

2 R.T. Watson, "Chemical Kinetic Data Survey VIII. Rate Constants of CℓO$_x$ of Atmospheric Interest", NBSIR 74-516 (Interim) June 1974.

3 K.D. Foster and G.H. Kimbell, Fourteenth International Symposium on Combustion, 1973, p 203.

Gas Phase Reactions of Tin and Germanium Atoms

J.R. Wiesenfeld[*] and M.J. Yuen

Department of Chemistry, Cornell University
Ithaca, NY 14853, USA

The chemistry of Group IV atoms in the presence of oxidizing gases is of considerable potential importance in the development of a visible chemical laser. Indeed, a high photon quantum yield has been observed [1] as the result of the reaction

$$Sn + N_2O \rightarrow SnO^* + N_2$$

where the production of electronically excited triplet SnO molecules may be predicted strictly on the basis of spin conservation, the ground state of the atom being $5\,^3P_J$. In the experimental program described here, the reaction kinetics of these oxidation reactions are examined in detail with special emphasis being placed on the elucidation of the role played by electronic structure in the course of the chemical reaction.

The experimental technique utilized in these experiments was similar to that previously applied to wide range of atom-molecule reactions, namely flash photolysis-atomic absorption spectroscopy. A sample of $SnCl_4$, $Sn(CH_3)_4$ or $GeCl_4$ was photolyzed using a standard white-light flash photolysis unit. Typical conditions were $P_{MX_4} = 5 \times 10^{-4}$ torr, $P_{Ar} = 30$ torr, $E = 400$–500 J. The large buffer/source ratio ensures that the reagents in the quartz reaction vessel remain adequately thermalized at the temperature of the experimental apparatus, 298°K. The Sn and Ge atoms were detected at very low partial pressures (ca. 10^{-5}–10^{-6} torr) by observing the attenuation of the atomic resonance lines $Ge(4p^2, {}^3P_0) \rightarrow Ge(5s, {}^1P_1{}^0)$ at 249.8 nm and $Sn(5p^2, {}^3P_0) \rightarrow Sn(6s, {}^1P_1{}^0)$ at 286.33 nm. The temporal profile of the intensity variation in the transmitted signal was monitored photoelectrically and displayed on an oscilloscope. Large signal/noise ratios permitted one-shot experiments and manual data analysis.

As reactant gases were added to the reaction vessel in vast excess of the atomic species, the removal of Ge and Sn was observed to be first-order with respect to atom concentrations. The measured rate coefficient, k, could be related to the bimolecular rate constant, k_{reac}, by

$$k = k_{reac}[\text{reactant}] + K$$

where K accounts for reaction with the source gas or impurities in the buffer gas.

The reaction rate constants for Ge and Sn were measured for a variety of oxidizers (Table 1). The wide variation in reaction efficiency is striking,

---

[*]Camille and Henry Dreyfus Foundation Teacher-Scholar

Table 1 Biomolecular rate constants for group IV oxidation reactions, T=295°K

| Oxidizer | Ge | | | Sn | | |
	k [cm³molec⁻¹sec⁻¹]	ΔH [eV]	cross-section [A°²]	k [cm³molec⁻¹sec⁻¹]	ΔH [eV]	cross-section [A°²]
Cl_2	6.1(-11)[a].	-1.0	14.5	5.3(-11)	-0.8	14.0
O_2	2.5(-10)	-1.7	46.8	1.1(-11)	-0.2	2.1
N_2O	9.9(-12)	-4.9	2.1	6.2(-16)[b].	-3.7	—
NO_2	4.6(-10)	-3.9	96.9	4.0(-10)	-2.3	91.4
COS	3.4(-10)	-2.0	77.4	1.3(-11)	-1.1	3.2

a. Numbers in parenthesis represent exponents.
b. Extrapolated from data presented in Reference 3.

with some reactions displaying essentially gas-kinetic cross-sections. Also, the rate of reaction of Ge with several of the oxidizers is much faster than that for Sn. Thus, a mechanism based on an "electron-jump" to account for the large cross-sections [2] must be considered unlikely, the first ionization potential of Ge being 0.79 eV higher than that of Sn. The exothermicities of the reactions under study are also given in Table 1. It is evident that the higher energies available to the Ge + oxidizer reactions may play a role in facilitating the enhancement of these processes. In particular, the reaction of Ge with N_2O which proceeds over 10^3x faster than the corresponding oxidation of Sn may involve a low activation barrier which permits relatively rapid reaction. The presence of such a barrier, induced by diabatic correlations of Sn + N_2O with high-lying states of SnO, has previously been discussed [3].

The reactions of Sn and Ge with O_2 may profitably be compared with that of C with O_2. Thrush and coworkers [4] demonstrated that the CO molecules so produced

$$C(^3P_J) + O_2 \rightarrow CO(v'' \leq 17) + O(^1D_2)$$

possessed a vibrational energy distribution expected for concomitant production of electronically excited oxygen atoms, $O(^1D_2)$, i.e., the highest vibrational energy observed would involve conversion of virtually all of the product energy into vibration. This observation was explained in terms of the need to maintain spin conservation of $C(^3P_J) + O_2$ via the low lying singlet surface of CO_2 which correlated with CO + $O(^1D_2)$ and not CO + $O(^3P_J)$. A similar argument would hold for Ge or Sn + N_2O. In these cases, however, production of MO + $O(^1D_2)$ would be endothermic with respect to reactants, suggesting that the proposed mechanism must be modified. Indeed, unpublished work of Herm and coworkers [5], indicates that the collision complex, $SnO_2^†$, resulting from Sn + O_2 is extremely long-lived. This, in turn, leads to the conclusion that the high efficiency for the reactions of Sn and Ge with O_2 is due to the formation of a singlet complex which is so stable that it passes through the region of crossing with the triplet surface correlating with CO + $O(^3P_J)$ many times, eventually undergoing a non-adiabatic transition to this repulsive surface which rapidly dissociates to product MO + $O(^3P_J)$.

Finally, we address the interesting behavior of Ge and Sn upon collision with COS. As may be inferred from Table 1 and the standard compilations of spectroscopic constants [6], the reaction

$$M(^3P_J) + COS \rightarrow CO + MS$$

is very efficient (especially for Ge) in spite of the inaccessibility of any known triplet states of GeS or SnS which would be required in order to conserve spin in this reaction. Thus, this appears to be a case where a nonadiabatic chemical reaction may be highly efficient. It is not at all clear why this process is so efficient for COS oxidation but not N_2O, which are valence shell isoelectronic with one another. The possible role played by metastable electronic states of the atoms as well as the nature of the product states are currently under investigation in this laboratory.

This work was supported by the United States Air Force Office of Scientific Research.

References

1. W. Felder and A. Fontijn: Chem. Phys. Letters, 34, 398(1975)

2. D.R. Herschbach: Advan. Chem. Phys., 10, 319(1966)

3. J.R. Wiesenfeld and M.J. Yuen: Chem. Phys. Letters, 42, 293(1976)

4. E.A. Ogryzlo, J.P. Reilly and B.A. Thrush: Chem. Phys. Letters, 23, 37 (1973)

5. R. Herm, private communication

6. S.N. Suchard: "Spectroscopic Constants for Selected Heteronuclear Diatomic Molecules", SAMSO-TR-74-82, Vol. I,II,III (1974)

The Possibility of Laser Pumping via Energy Transfer from or Reactive Collisions with I ($^2P_{1/2}$)

P.L. Houston

Department of Chemistry, Cornell University
Ithaca, NY 14853, USA

Collisions of electronically excited iodine atoms I^* ($*=^2P_{1/2}$) can result in efficient energy transfer to vibrational modes of the collision partner (E-V transfer) or in reactions which produce electronically and/-or vibrationally excited products. Either of these processes might form the pumping scheme for a new laser system. This paper will outline our recent progress in understanding these effects.

Vibrational fluorescence has been observed from HCl, HBr, NO, and H_2O following energy transfer from I^* [1,2]. The experimental apparatus employed for these observations is shown in Figure 1. A flash-lamp pumped dye laser was used to dissociate I_2 at 490 nm to yield

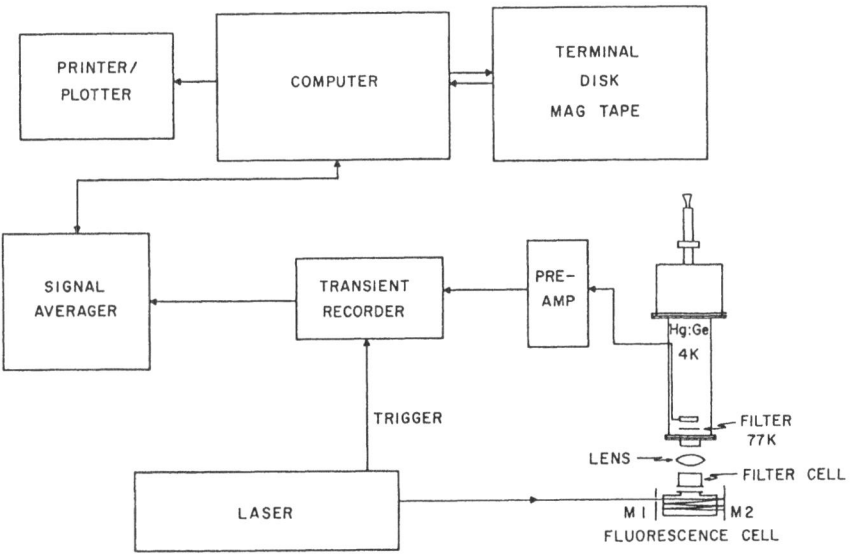

Fig. 1 Experimental Apparatus for Investigation of I^* Energy Transfer

I^* + I. Infrared fluorescence from either the electronically excited I^* (1.315 μ) or the vibrationally excited collision partner was collected with a lens, separated by a filter, and detected by an infrared sensitive element (Hg:Ge or In:Sb). Signals were digitized with a transient recorder, averaged in a hard-wired averager, and sent to a computer for analysis. A typical vibrational fluorescence trace is shown in Fig. 2.

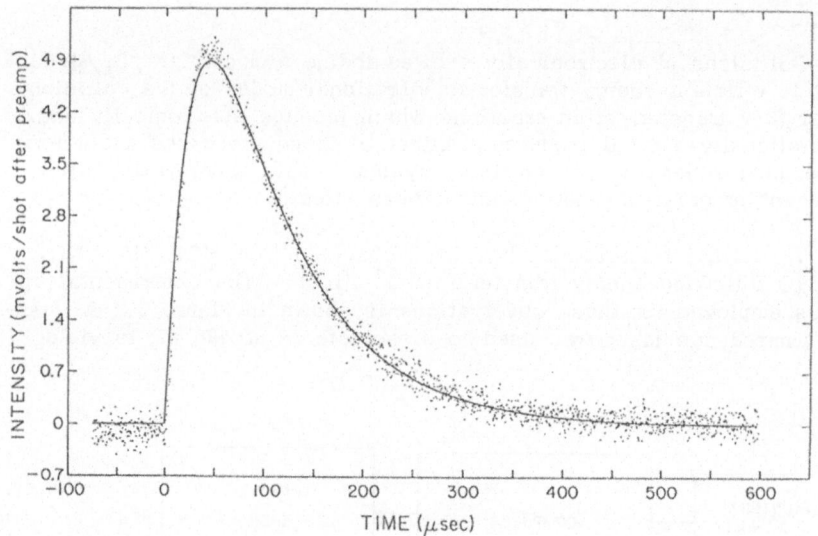

Fig. 2 Averaged Fluorescence Trace for 4.30 torr of NO

Fluorescence traces were analyzed according to the following kinetic scheme:

$$I^* + AB(v=0) \xrightarrow{k_i^E} I + AB(v=i) \tag{1}$$

$$AB(v=i) + AB(v=0) \xrightarrow{k_{vv}} AB(v=i-1) + AB(v=1) \tag{2}$$

$$AB(v=i) + M \xrightarrow{k_{i,i-1}^V} AB(v=i-1) + M \tag{3}$$

where AB = HCl, HBr, NO, or H_2O and M = AB, I_2, or argon. The decay of fluorescence from I^* or the rise in fluorescence from AB yields the total deactivation rate for I^* by AB, Σk_i^E. These values, along with recent literature determinations, are listed in Table I. Comparison of the amplitudes of the I^* and AB fluorescence should yield information concerning the fraction of electronic energy transferred into vibrational energy. This work is now in progress.

The table also lists the vibration-to-translation (V-T) rates found in this study as well as some literature values for the vibration-to-vibration (V-V) relaxation rates. It should be noted that when AB is HCl, HBr, or NO, the V-V rate is always faster than Σk_i^E. Consequently, it is unlikely that these systems will produce inversion in the

Table 1: Observed Deactivation Rates

AB	Rate	I^*	AB	Lit.	
			$sec^{-1} \, torr^{-1}$		
HCl	Σk_i^E	$(4.46 \pm .77) \times 10^2$		$(4.89 \pm .39) \times 10^2$	[3]
				$(2.09 \pm .29) \times 10^2$	[4]
	k_{10}^V			$(8.3 \pm .8) \times 10^2$	[5]
	k_{21}^{VV}			$(1.0 \quad) \times 10^5$	[6]
HBr	Σk_i^E	$(3.51 \pm .12) \times 10^3$	$(3.64 \pm .46) \times 10^3$	$(4.19 \pm .32) \times 10^3$	[3]
				$(5.09 \pm .23) \times 10^3$	[4]
	k_{10}^V		$(6.0 \pm .9) \times 10^2$	$(5.71 \pm .5) \times 10^2$	[7]
	k_{21}^{VV}			$(1.4 \quad) \times 10^5$	[8]
NO	Σk_i^E	$(3.88 \pm .32) \times 10^3$	$(3.84 \pm .34) \times 10^3$	$(5.1 \quad) \times 10^4$	[9]
	k_{10}^V		$(2.51 \pm .18) \times 10^3$	$(2.5 \pm .15) \times 10^3$	[10]
	k_{21}^{VV}			$(1.1 \pm .24) \times 10^5$	[10]
H_2O	k_i^E	$(6.73 \pm .99) \times 10^4$	$(7.32 \pm .88) \times 10^4$	$(7.41 \pm .97) \times 10^5$	[11]

diatom. However, for H_2O it may be possible that selective E-V and V-V pumping will produce an inversion between bending combination modes.

The rate of reaction of I^* with Br_2 has also been determined [12]. Fig. 3 shows the variation in I^* fluorescence deactivation rate vs. the mole fraction of Br_2. From the slope of the straight line, it may be determined that the total deactivation rate of I^* by Br_2 is extremely fast, $k = (1.95 \pm .12) \times 10^6 \, sec^{-1} \, torr^{-1}$. This total deactivation is the sum of the rates for three processes:

$$I^* + Br_2 \xrightarrow{k^*} IBr + Br^* \qquad (4)$$
$$I^* + Br_2 \xrightarrow{k} IBr + Br \qquad (5)$$
$$I^* + Br_2 \xrightarrow{k_d} I + Br_2 \qquad (6)$$

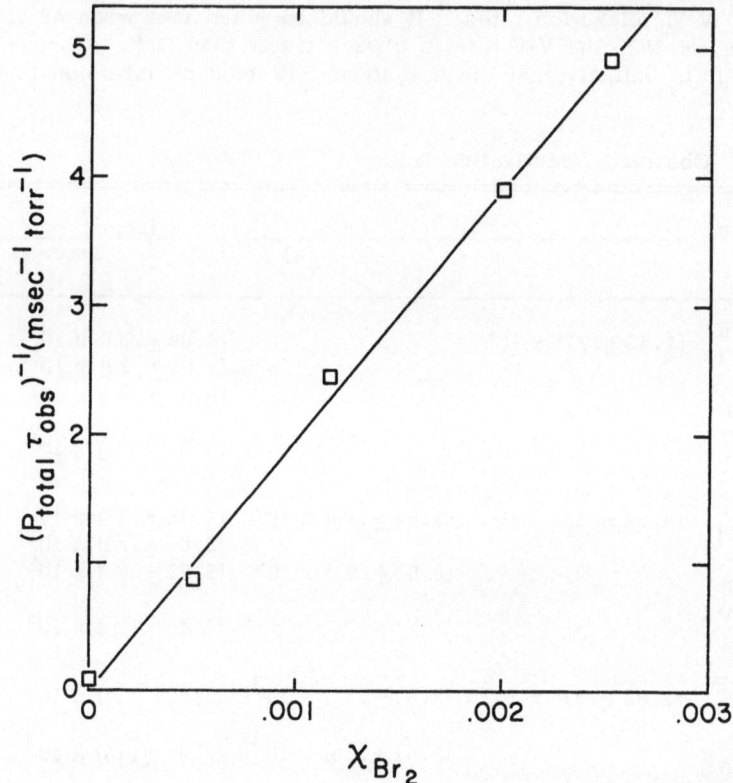

<u>Fig. 3</u> I* Fluorescence Decay Rate as a Function of Mole Fraction of Br$_2$

A correlation diagram for the IBrBr system is shown in Figure 4. I* and Br$_2$ react adiabatically on a 2^2A' surface to yield IBr and Br*. The reaction of I to form IBr and Br is adiabatically allowed by thermally inaccessible. Consequently, the reaction of I* with Br$_2$ may produce inversion on the Br*-Br transition even in the absence of I* inversion. This possibility is currently under investigation in our laboratory.

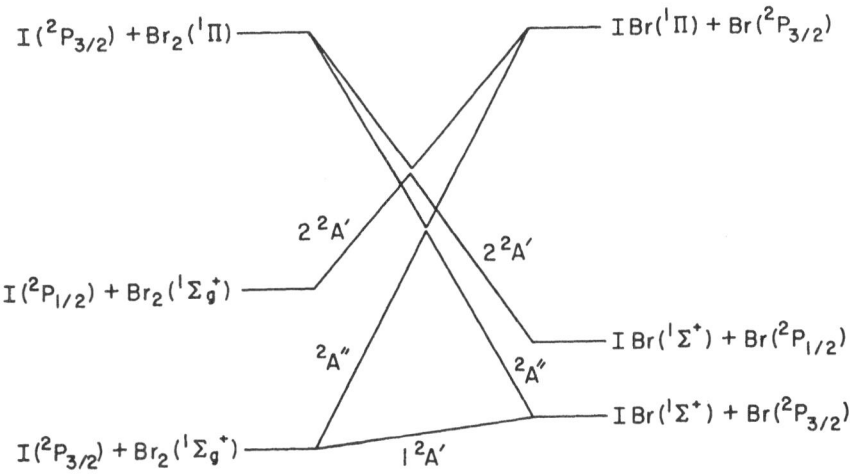

$I(^2P_{3/2}) + Br_2(^1\Pi)$

$IBr(^1\Pi) + Br(^2P_{3/2})$

$2\,^2A'$

$2\,^2A'$

$I(^2P_{1/2}) + Br_2(^1\Sigma_g^+)$

$^2A''$

$^2A''$

$IBr(^1\Sigma^+) + Br(^2P_{1/2})$

$IBr(^1\Sigma^+) + Br(^2P_{3/2})$

$I(^2P_{3/2}) + Br_2(^1\Sigma_g^+)$

$1\,^2A'$

Fig. 4 Correlation Diagram for the IBrBr System

References

1. A.J. Grimley and P.L. Houston, to be published.
2. P.L. Houston and A.J. Grimley, in Electronic Transition Lasers II, Wilson, Suchard and Steinfeld, eds., p 257-265.
3. R.J. Donovan, C. Fotakis, and M.F. Golde, J. Chem. Soc. Farad. Trans. II, 72, 2055 (1976).
4. A.T. Pritt and R.D. Coombe, J. Chem. Phys. 65, 2096 (1976).
5. H.-L. Chen and C.B. Moore, J. Chem. Phys. 54, 4072 (1971).
6. B. Hopkins and H.-L. Chen, J. Chem. Phys. 57, 3816 (1972).
7. H.-L. Chen, J. Chem. Phys. 55, 5551 (1971).
8. B. Hopkins and H.-L. Chen, Chem. Phys. Lettr. 17, 500 (1972).
9. J.J. Deakin and D. Hussain, J. Chem. Soc. Farad. Trans. II, 68, 41 (1972).
10. J. Stephenson, J. Chem. Phys. 59, 1523 (1973).
11. D. Burde and R. McFarlane, J. Chem. Phys. 64, 1850 (1976).
12. P.L. Houston, Chem. Phys. Lettr. 47, 137 (1977).

Part III

Other Laser Systems

Ultraviolet Ion Lasers

J.R. McNeil, R.D. Reid, D.C. Gerstenberger, and G.J. Collins

Department of Electrical Engineering, Colorado State University
Fort Collins, CO 80523, USA

Abstract

We have extended cw laser action down to 220 nm. This represents a 100 nm
improvement to the state-of-the-art prior to our work. Moreover, we have obtained cw laser oscillation on twenty additional transitions in the spectral
region between 220 and 320 nm.

The successful laser systems, dominant laser transitions and maximum output
power may be summarized as follows: Ne-Cu at 260 nm (350 mW); Ne-Ag at 318 nm
(>350 mW) and He-Au at 280 nm (100 mW). It is noteworthy that ultraviolet
laser threshold currents as low as 2 A have been observed. In comparison rare
gas ion lasers require 20 - 50 A to reach threshold. Finally, the metals of
interest are sputtered into the discharge allowing us to obtain metal densities
of 10^{14} atoms/cm^3 without the use of external ovens or discharge heating.

Our most recent attempts to scale the output power levels to one watt cw will
be presented with particular emphasis on the 780 nm transition of Cu II.

1. Introduction

1.1 Past Work

The primary sources of cw ultraviolet laser action are the Kr^{++}, Ar^{++} and Cd^+
lasers which operate continuously at 364, 351 and 325 nm respectively. We
have extended the range of cw laser oscillation down to 220 nm in the work
described below as well as demonstrating that one can obtain 350 mW of output
power at 260 and 318 nm. Finally, our attempts at obtaining a one watt ultraviolet laser operating below 320 nm will be described.

1.2 Charge Transfer Lasers

Charge transfer reactions of the form,

$$R^+ + M \rightarrow (M^+)^* + R + \Delta E \text{ (kinetic energy)} \qquad \text{Reaction (1)}$$

have been investigated for a possible ultraviolet laser, where R^+ is a ground
state helium or neon ion, M^{+*} is an excited metal (Cu, Ag or Au) ion and ΔE
is the energy differences between R^+ and M^{+*}. We have already shown in pre-

vious work that charge transfer is a favorable pumping scheme for pumping ultraviolet laser transitions because: cross-sections for charge-transfer reactions [1,2] are in excess of $10^{-15} cm^2$, the sharp selectivity of charge-transfer reactions for forming upper laser levels but *not* lower levels [3]; the capability of creating ion densities in excess of 10^{15} ions/cm^3 in high-pressure gas discharges excited by an electron beam; and finally, laser oscillation occurs in the singly ionized spectrum rather than in multiply ionized species, resulting in higher quantum efficiency.

We believe that ultraviolet lasers using charge-transfer collisions as a pumping scheme have the potential for total system efficiencies approaching 0.2%. In total system efficiency, we include the following: the discharge efficiency for creating ground-state rare-gas ions; the efficiency with which we transfer this energy to the excited metal ion via a collision such as (1); and the quantum efficiency of the laser transition itself. The quantum efficiency for a He^+ -, Ne^+ -, Ar^+ -, or Xe^+ - pumped ion laser operating at 2500 Å is 20, 24, 36, and 41%, respectively. Only rare-gas-halogen lasers have larger quantum efficiencies and larger demonstrated operating efficiencies.

The hollow cathode discharge is particularly simple to model for the charge-transfer mechanism. To a first approximation, the efficiency of creating rare-gas ions in a hollow cathode discharge is given by $\gamma/(1+\gamma)\beta$, where γ is the second Townsend coefficient of the cathode material and βV_i is the average energy lost by a beam electron emerging from the cathode dark space [4]. The over-all efficiency for the charge transfer laser, η, can then be expressed as,

$$[\gamma/(1 + \gamma)\beta](V^*/V_i) = \eta_i \tag{1}$$

where V^*V_i is the quantum efficiency, V^* is the photon energy, and V_i is the ionization potential of the rare gas. For the Ne-Cu laser system the calculated total system efficiency is 0.2%.

Outlined below are results to date.

2. Summary of Results

2.1 Ne-Cu Laser

We have obtained cw laser oscillation on the seven Cu II transitions shown in Fig. 1. The laser transitions were excited in a copper hollow cathode described below. Note in Fig. 1 that the ^1D and ^3D terms of the $3d^95s$ electronic configurations of Cu II give rise to all seven transitions. Both the ^1D and the ^3D terms lie in near energy coincidence to the ground state neon ion and a charge-transfer reaction,

$$Ne^+ + Cu \rightarrow (Cu^+)^{**} + Ne + \Delta E \quad , \qquad \text{Reaction (2)}$$

is thought to provide the dominant excitation mechanism. We have obtained 350 mW of output power from the 260.0 and 299 nm transitions in our preliminary work and will be attempting one watt output.

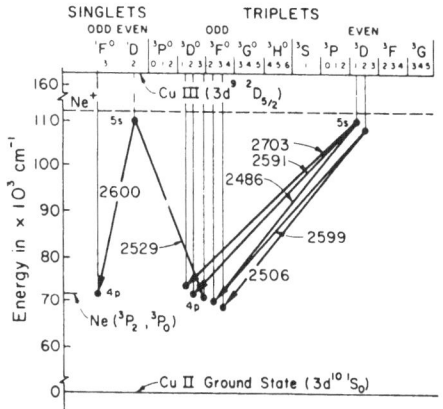

Fig. 1. A partial term diagram of Cu II indicating the seven ultraviolet laser wavelengths and transition assignments. The energy available from Ne^+ and Ne^M ($^3P_2$, $^3P_0$) in a collision of the second kind is also shown

Table 1. New Metal Ion Laser Transitions

Air [Å]	Ion	Power Output [mW]
2243	Ag	0.1
2278	Ag	0.1
2485	Cu	2
2506	Cu	2
2529	Cu	1
2590	Cu	100
2599	Cu	100
2600	Cu	150
2703	Cu	200
2822	Au	30
2847	Au	50
2894	Au	50
2918	Au	50
3181	Ag	350
4086	Ag	250
4788	Ag	30
7404	Cu	50
7556	Au	50
7600	Au	50
7739	Cu	30
7808	Cu	1000
7826	Cu	20
7896	Cu	50
7944	Cu	50
7988	Cu	50
8004	Ag	100
8404	Ag	1000

2.2 Ne-Ag Laser

Table I lists the ten transitions of Ag II which have been obtained when a silver hollow cathode is excited with a helium or neon buffer gas. Note that previously reported *unidentified* laser transitions [5,6] are now known to

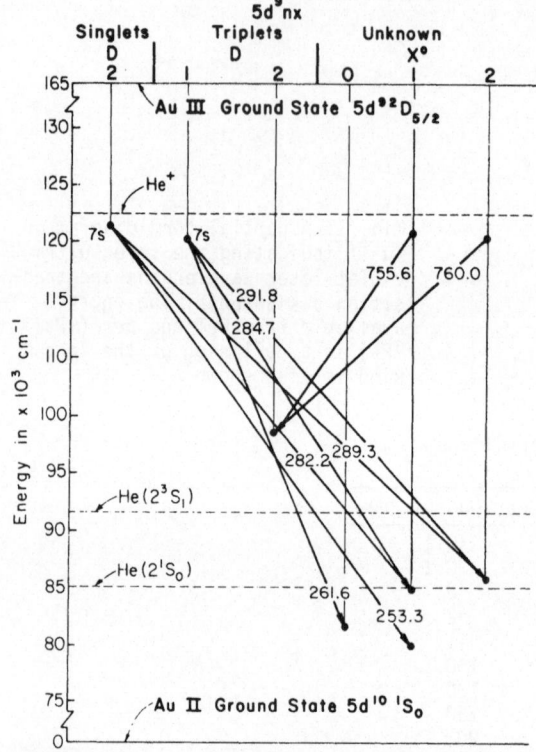

Fig. 2. Partial term diagram of Au II with representative ir and uv laser transitions (solid lines). Selected energy levels from the 4d³nx electronic configuration are shown and the energy available from the ground-state helium ion, He⁺, in a thermal energy charge-transfer collision. Reaction (1) is also shown. All wavelengths are in nm

arise from grating ghosts [7]. Details of the laser tube construction are given in section D. The 227 and 224 nm transitions, $5d^3D_2 - 5p^3P^0_1$ and $5d^1S_0 - 5p^1P^0_1$ respectively, are the shortest wavelength cw laser transitions reported in the literature to date. The strongest laser transition of Ag II is at 318 nm, $4d^85s^2 \, ^1G_4 - 4d^95p \, ^3F^0_3$, provides peak single-line output power of 350 mW. We are presently designing silver hollow cathodes with the goal of one-watt output at 318 nm.

2.3 He-Au Laser

Figure 2 indicates the six ultraviolet laser transitions of Au II which we have observed when a helium discharge is excited in a gold hollow cathode. Note that all of the Au II laser transitions arise from energy levels in near energy coincidence with the ground state helium ion. Threshold currents for the 280 nm laser transitions are measured to be as low as 3A or about a factor of twenty less than threshold currents for ultraviolet laser transitions in rare-gas-ion lasers. Multi-line output power of 125 mW has been demonstrated in the 250 to 290 nm region.

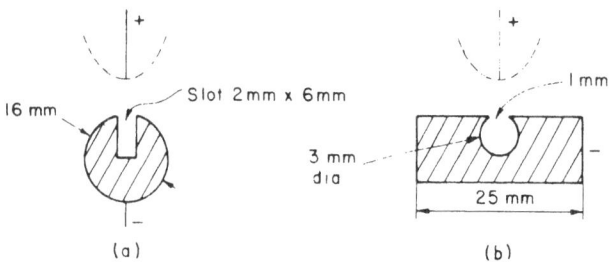

Fig. 3a and b. Cross sections of the hollow cathode geometries employed by:
(a) rectangular aperture and (b) circular aperture. The internal surface of
both cathodes is coated with gold of thickness 0.1 nm. Note that the anode
is constructed of stainless-steel mesh to reduce discharge instabilities

2.4 Hollow Cathode Laser Tube

The hollow cathode configurations employed in our Cu II, Ag II and Au II
studies are shown in Fig. 3. In all systems the hollow cathodes were 50 cm
long. It is noteworthy that the metal vapor of interest was created via discharge sputtering rather than via an external oven or via discharge heating.
Measurements of the ground state copper density are summarized in Fig. 4.
Under optimum conditions a copper density of 10^{14} atoms/cm^3 was created with
the cathode at 20°C. Hence, our vacuum envelop was made of pyrex and high
temperature materials were *not* required. We have experimented with a variety of anode and cathode geometries and have achieved laser action from circular as well as rectangular cross-section slots. It is noteworthy that TEM..
output is best achieved in the cylindrical configuration, although ease of
construction favors a rectangular slot if transverse mode purity is not important. Also shown in Fig. 3 is the anode geometry employed. The anode consisted of a stainless-steel mesh placed 5mm above the cathode slot, extending
the entire length of the cathode. We found that the mesh anode allows higher
discharge currents to be employed without arc formation as compared to other
anode configurations. The curve-shaped anode illustrated in Fig. 3 gives the
best performance, to date, of all anode geometries we have attempted.

3. Relation of this Work to Other Ultraviolet Lasers

3.1 Other Ultraviolet Sources

Table II summarizes all gas laser sources operating in the ultraviolet. Clearly, the rare-gas-halogen lasers offer a high power pulsed source of ultraviolet radiation that will have a direct application to practical isotope
separation schemes.

The Cu$^+$ and Ag$^+$ ion lasers developed at Colorado State University, on the
other hand, offer very clear and distinct advantages over previous rare-gas-
ion lasers. For the first time cw laser sources are available below 320 nm
and at power levels approaching those of the rare-gas-ion lasers. The Ag$^+$
and Cu$^+$ lasers are only one year old while the rare-gas-ion lasers have been

Table 2. Ultraviolet lasers

	Continuous Wave Ion Lasers				Pulsed Molecular Lasers			
Laser	Ar^{++}, Kr^{++}	Cd^+	Cu^+	Ag^+	N_2 Gas discharge	ArF E-Beam	KrF Gas discharge	XeF Gas discharge
Wavelength [Å]	3511-3638 3507-3564	3250	2486-2703	2243-3831	3159 3371 3577	1933	2484-2491	3511 3531
Threshold Current [A]	40	0.1	6	7	10^4	10^4	10^4	10^4
Voltage [V]	500	1000	260	280	10^4	10^6	10^4	10^4
Power out [W]	2-5	0.01	0.35	0.36	10^5	10^9	10^4	10^6
Quantum Efficiency [%]	5	21	24	24	45	50	50	50

developed for nearly fourteen years. Clearly our goal of one watt cw in the near future is a real one. It is not unrealistic to envision multi-watt output.

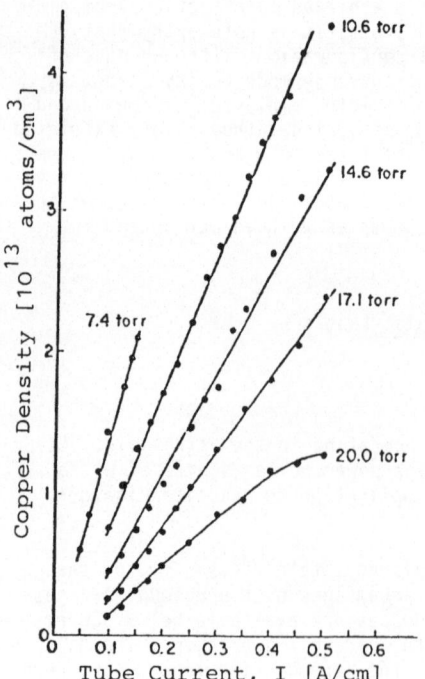

Fig. 4. Copper density versus discharge current

3.2 Application to Tunable Dye Laser (300 - 400 nm)

One major reason to develop a one watt ultraviolet laser is so that one can pump ultraviolet dyes and obtain TUNABLE cw laser action in the 300 - 400 nm region. Present dye laser technology cannot go below 415 nm on a cw basis. Because of its tunability and its spectrally pure output, the cw dye laser is a more valuable tool for many applications in photo-chemistry and spectroscopy than a pulsed laser because: a narrower spectral output is available, potentially higher average power is available, and because cw output makes possible much simpler detection electronics as compared to pulsed nanosecond detection electronics which is typical of pulsed dye laser operations.

References

1 T. Shay, H. Kano, G.J. Collins: Appl. Phys. Lett. 26, 531 (1975)
2 H. Kano, T. Shay, G.J. Collins: Appl. Phys. Lett. 27, 610 (1975)
3 G.J. Collins: J. Appl. Phys. 44, 4633 (1973); 46, 1412 (1975)
4 A. Von Engel: *Ionized Gases* (Oxford University Press, London 1965)
5 W.L. Johnson, J.R. McNeil, G.J. Collins, K.B. Persson: Appl. Phys. Lett. 29, 101 (1976)
6 J.R. McNeil, W.L. Johnson, G.J. Collins, K.B. Persson: Appl. Phys. Lett. 29, 172 (1976)
7 R.D. Reid, D.C. Gerstenberger, J.R. McNeil, G.J. Collins: J. Appl. Phys., Oct. (1977)

CW Optically Pumped Molecular Iodine Laser[1]

J.B. Koffend, F.J. Wodarczyk[2], and R.W. Field[3]

Department of Chemistry, Massachusetts Institute of Technology
Cambridge, MA 02139, USA

1. Introduction

Laser oscillation on several molecular iodine $B0_u^+ - X^1\Sigma_g^+$ transitions is
observed. The unique feature of this laser is that it is excited by
a continuous wave optical pump, a single longitudinal mode Ar$^+$ laser,
and that it exhibits cw oscillation on more than 100 individually selectable transitions spanning more than one octave from 0.57 to 1.35μ.
Optimum gain, pump threshold, and photon conversion efficiency, respectively 0.004 cm^{-1}, 4 mW, and 10%, are obtained for laser transitions
near 1.3μ.

1.1 Other Optically Pumped Electronic Transition Lasers

Interest in optically pumped lasers (OPL) is growing rapidly. Optically
pumped vibrational lasers, operating at infrared wavelengths from 10-
1800μ have been well known since 1970 [1] and have found numerous
spectroscopic and device applications. Photon conversion efficiencies
(photon out per absorbed pump photon) in excess of 15% have been achieved for pulsed vibrational OPLs [1]. The development of electronic
transition OPLs is summarized in Table 1.

It is becoming increasingly clear that, given a suitable pump laser,
virtually any diatomic molecular transition may be made to lase. This
includes electronic transitions that are unobservably weak in spontaneous
fluorescence. With a tunable pump laser, each electronic transition
contains as many as 10^6 potential laser lines. Although it appears
to be easier to obtain high photon conversion efficiencies from pulsed
OPLs, a 50% efficient single line cw OPL is beginning to appear plausi-
ble.

With the exception of the HgBr laser, which is really a photodissociation
rather than optically pumped laser, no heteronuclear diatomic
OPLs have been reported. This is an accidental situation to which no

1. This research was supported by a grant from the Air Force Office
   of Scientific Research AFOSR-76-3056.

2. Rome Air Development Center (ESO)
   Hanscom Air Force Base, Massachusetts 01731

3. Alfred P. Sloan Fellow.

Table 1   Optically pumped electronic transition lasers

Transition	Pumped Laser	Year	Remarks	Reference
$I_2 B0_u^+ - X^1\Sigma_g^+$	doubled Nd	1972	pulsed	[2]
$Na_2 B^1\Pi_u - X^1\Sigma_g^+$	doubled Nd	1975	pulsed	[3]
$Na_2 A^1\Sigma_u^+ - X^1\Sigma_g^+$	doubled Nd	1975	pulsed	[3]
	flashlamp dye	1975	pulsed	[4]
$Br_2 B0_u^+ - X^1\Sigma_g^+$	doubled Nd	1976	pulsed	[5]
$S_2 B^3\Sigma_u^- - X^3\Sigma_g^-$	doubled dye	1976	pulsed	[6]
$I_2 B0_u^+ - X^1\Sigma_g^+$	argon ion	1976	cw	[7]
$Na_2 B^1\Pi_u - X^1\Sigma_g^+$	argon ion	1976	cw	[8]
$Te_2 A0_u^+ - X0_g^+$	flashlamp dye	1977	pulsed	[9]
$Li_2 B^1\Pi_u - X^1\Sigma_g^+$	argon ion	1977	cw	[10]
$HgBr \, B^2\Sigma^+ - X^2\Sigma^+$	ArF	1977	pulsed HgBr$_2$ photodissociation	[11]
$I_2 B0_u^+ - X^1\Sigma_g^+$	flashlamp dye	1977	pulsed	[12]

practical or physical significance should be attached. No triatomic
or small polyatomic molecule which is not considered a dye has lased on
an electronic transition. Polyatomic OPLs face the more fundamental
limitations of dilution of optical pump power among many unresolvable
transitions and photochemical quenching of excited species.

1.2 Difference Between Electronic and Vibrational OPLs

The integrated cross-section, $\sim \sigma° \, \Delta\nu$ ($\sigma°$ is the cross-section at line
center and $\Delta\nu$ is the linewidth), for vibrational transitions is seldom
larger than 10^{-15} cm^2 whereas the maximum value for electronic transitions
is 10^{-12} cm^2. Thus electronic transition OPLs should be capable of
higher gain per mode volume, lower pump threshold, better conversion
efficiency, and should be operable at much higher pressure without
significant quenching. The larger cross-section means that it is easier
to power broaden the pump transition, thus facilitating short mode-locked
pulses throughout the visible and infrared spectral regions. Finally, a
much wider variety of transitions are accessible from an electronically
excited level; these might include several different electronic transitions, pure vibrational or rotational transitions, and a larger range
of lower vibrational levels than are permitted by pure vibrational
selection rules. Chemically and spectroscopically interesting lower
levels, particularly those near dissociation limits, might be selectively
populated.

98

1.3 Applications of OPLs

OPLs will have an impact on three major areas of research: spectroscopy,
kinetics, and new laser devices. Transitions observable as spontaneous
laser induced fluorescence might be observed as self-stimulated laser
transitions with a tremendous solid-angle detection advantage. Unobservably weak spontaneous transitions have been made to lase and the
spectrum of laser output can be freed of both collision induced rotational
satellites as well as stronger interfering transitions. Collisional
transfer rates out of upper and lower laser levels are sampled by the
time, pump power, and pressure dependence of OPL output. More importantly,
in the laser gain medium itself various selectable, vibrationally or
electronically excited rovibronic lower laser levels (including metastable "reservoir states") may be significantly populated in a
pulsed or cw manner. The absolute population flux into such levels is
signalled by the laser output power, while the rates of collisional
population of neighboring levels might be sampled either in the side-
fluorescence or with a tunable probe laser. Practical OPL applications include frequency standards, multi-octave line tunable devices,
gain diagnostics for new chemical or electric discharge lasers,
frequency up and down conversion with improvement of spatial mode
quality, and generation of subnanosecond mode locked pulses.

2. The Ar^+ Laser Pumped CW I_2 B-X Laser

A 50 cm Brewster cell of I_2 vapor at room temperature ($\sim$ 0.3 torr, no
buffer gas) is pumped with a Spectra Physics 171-00 Ar^+ laser operated
in a single longitudinal mode with an intracavity etalon. Power available at 514.5 nm is 8W multimode and 4W single mode. The experimental
configuration is illustrated by fig. 1.

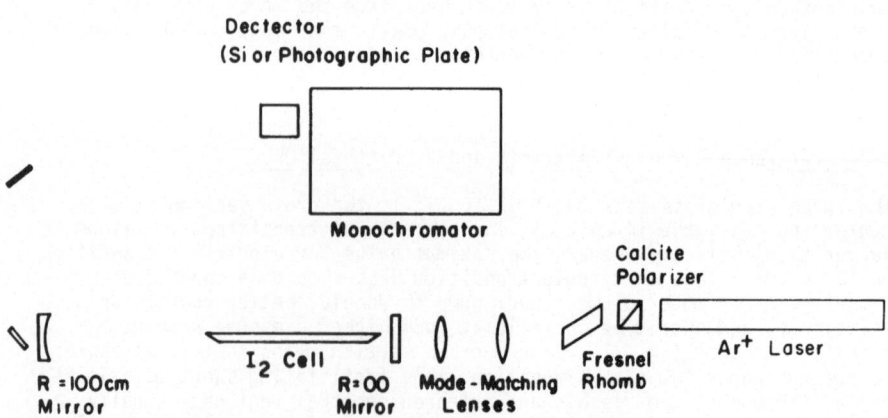

Fig. 1 Diagram of optically pumped I_2 laser. Incident Ar^+ pump radiation
is circularly polarized. The pump waist is located at the R = ∞ mirror
in the plano-spherical configuration shown, or at the midpoint of the
symmetrical spherical cavity used for the 1.3μ I_2 transitions. At wavelengths beyond 1.1μ, I_2 laser output is detected with either a lead
sulfide or thermopile detector.

The most important features are the optical isolator, which is needed to reduce feedback into the Ar^+ laser due to back-reflected 514.5 nm radiation from the I_2 resonator and input optics, and the mode matching lenses. It is essential to superimpose the pump beam onto the TEM_{00} mode of the I_2 resonator; otherwise pump power would be wasted exciting I_2 molecules which cannot contribute to gain. Two mode matching lenses are used because in some diagnostic experiments, a high speed chopper is placed between these lenses at the Ar^+ beam waist.

Various cavity configurations have been employed. Fig. 1 shows the first (but least effective) of these, with which lasing was initially achieved for most of the transitions shown in fig. 2.

I_2 Potentials

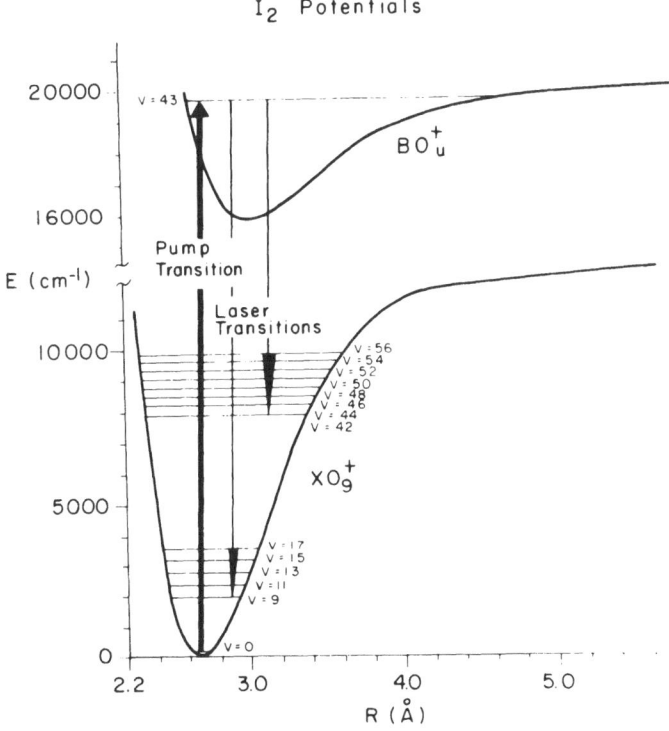

Fig. 2 RKR potential energy curves for the I_2 $B0_u^+$ and $X^1\Sigma_g^+$ states. The strongest pump transition and two of the three groups of laser transitions observed are indicated. The 1.3μ transitions are not shown. The observed transitions are restricted to regions of high mirror reflectivity.

The pump laser enters the 95 cm long I_2 plano-spherical (R = 100 cm) resonator through the plane mirror. The mirrors transmit more than 85% at 514.5 nm and have greater than 98% reflectance in the regions 570-970 nm or 640-1040 nm. A much more successful configuration utilizes two R = 100 cm mirrors (0.15% transmission at 1.3μ) spaced approximately 100 cm apart. This permits the pump laser to be slightly skewed with respect to yet entirely contained within the I_2 resonator mode. Ar^+ laser feedback and burning of the plane mirror and input Brewster window are virtually eliminated.

I_2 Laser Emission in (43,56) Band

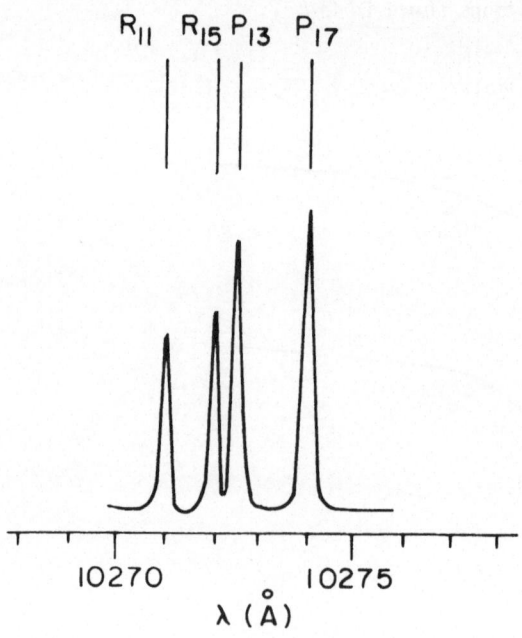

Fig. 3 Spectrum of the I_2 laser emission in the B-X (v' = 43, v" = 56) band. Four rotational lines appear because the 514.5 nm pump laser is tuned to excite accidentally coincident transitions, P(13) and R(15). The relative intensities of rotational lines are not determined by Hönl-London factors. No collisional satellites are observed to lase.

The Ar^+ laser is tuned to excite two accidentally coincident transitions of the I_2 B-X system, (43,0) P(13) and R(15), thereby populating BO_u^+ v' = 43, J' = 12 and 16. For each mirror set, laser transitions in several (v' = 43, v") bands are observed, each consisting of as many as four rotational lines: P(13) and R(11) from J' = 12; P(17) and R(15) from J' = 16. Figure 3 is a fully resolved spectrum of the I_2 laser in the (43,86) band. Observed transitions range from (43,9) at 570 nm to (43,83) at 1.34μ. Oscillation may be restricted to a single vibrational band (at wavelengths shorter than 1μ) by insertion of either a 0.015 inch thick birefringent quartz plate at Brewster's angle or an SF4 glass Brewster prism in a Littrow arrangement. The birefringent plate renders the isolator ineffective and does not tune beyond 1.0μ, but is trivial to align and less lossy than the prism.

2.1 I_2 Laser Characteristics

The operating characteristics of an OPL are determined by the following:

1. The pump power that can be inserted into a single pump transition within the mode volume of the OPL resonator. This, combined with the mode volume (~ 0.4 cm^3 for the I_2 laser) and the spontaneous fluorescence, predissociative and collisional loss rates from the upper level ($1/\tau = 1.5 \times 10^7$ sec^{-1} for I_2 BO$_u^+$ $v' = 43$ at 0.3 torr), determines the upper laser level population density $N_{v',j'}$, and, assuming an empty lower level, the maximum inversion density ΔN. If 1W of 514.5 nm pump radiation is inserted into a single I_2 transition the inversion density is $\Delta N = 4.3 \times 10^{11}$ cm^{-3}.

2. The stimulated emission cross-section. This is affected by pump intensity and bandwidth, pressure, and the circulating power in the OPL. A single mode pump laser excites a single velocity group and the stimulated emission cross-section is that for a homogeneously rather than Doppler broadened line. It is easy to power or pressure broaden such a narrow line, thus reducing the peak cross-section. For the I_2 B-X (43,56) transition at 1.03μ (Franck-Condon factor = 0.0046, transition moment squared = 1 Debye^2, linewidth (FWHM) = 4.8×10^6 Hz, $\sigma^\circ = 3.7 \times 10^{-14}$ cm^2 at low pump intensity.

3. Intracavity loss processes. These include output coupling, absorption from the upper laser level or partially relaxed lower levels, thermal lensing and diffraction (due to a pump volume significantly smaller than the OPL mode volume), and population bottlenecking in the lower laser level. The maximum fraction of the optically pumped $N_{v',j'}$ population which is available as laser output is the ratio of the stimulated emission rate to the total loss rate from the upper laser level. By monitoring the depth of modulation of side fluorescence upon blocking the rear laser mirror, this ratio is observed (with 2W of inserted pump power) to be at least 25% for I_2 transitions near 1.3μ.

2.1.1 Linewidth Calculation

The simplest possible picture of the optically pumped I_2 laser will be adopted in order to make factor of two estimates of gain-related parameters. Upper and lower bounds to the power broadening contribution to the linewidth are obtained by ignoring the variation of pump intensity within the I_2 gain cell.

All experiments were conducted at room temperature in pure I_2 vapor at 0.3 torr. The Doppler width is

$$\Delta\nu^D_{FWHM} = 4.7 \times 10^8 \text{ Hz}$$

and the homogeneous width of I_2 BO$_u^+$ $v = 43$ [13,14,15] is

$$\Delta\nu^H_{FWHM} = 1/\pi\tau \tag{1}$$

$$1/\tau = 1/\tau_{spontaneous} + 1/\tau_{predissociation} + 1/\tau_{collision}$$

$$= 2.3 \times 10^5 + 1.2 \times 10^5 + 1.5 \times 10^7 \qquad (2)$$

$$\Delta v_{FWHM}^{H} = 4.8 \times 10^6 \text{ Hz}.$$

The largest contribution to Δv^H is due to I_2 self-quenching [14]. An upper bound for the power broadening contribution to Δv^H is

$$\Delta v_{FWHM}^{POWER} \leq \frac{\mu E}{h} \qquad (3)$$

The transition moment for the pump transition

$$\mu = R_e \left[S_{J'J''} \, q_{v'v''} \right]^{1/2} = 9.4 \times 10^{-2} \text{ Debye} \qquad (4)$$

is given by the product of electronic (1 Debye), rotational (S = 1/2) and vibrational ($q = 1.8 \times 10^{-2}$ [16]) intensity factors, and the optical electric field

$$E \leq \frac{21.9 P^{1/2}}{\omega_0} = 8.4 \times 10^2 \text{ V/cm} \qquad (5)$$

is bounded by the incident pump intensity and the pump beam radius at its waist inside the I_2 cell

$$P = 1 \text{ Watt}, \ \omega_0 = 2.6 \times 10^{-2} \text{ cm}.$$

Thus

$$\Delta v_{FWHM}^{POWER} \leq 4.0 \times 10^7 \text{ Hz}.$$

It would require incident power of 140 W to cause $\Delta v^P \sim \Delta v^D$.

A lower bound to Δv^{POWER} is obtained from the ratio

$$\frac{N_{v'J'}}{N_{v''J''}^0} \leq \frac{N_{v'J'}}{N_{v''J''}} \leq \frac{\Delta v^H}{\Delta v^D} \qquad (6)$$

where the single and double primed quantum numbers refer respectively to upper and lower levels of the pumped transition and the super zero denotes the Boltzmann population density in the absence of pump radiation. Optical pumping may deplete one velocity group of the v",J" level unless collisional repopulation is sufficiently rapid. Since, on a time scale of 10^{-6} sec, the pump laser has a spectral width of less than 1 MHz, it can interact only with the fraction $\Delta v^H/\Delta v^D$ of the molecules in the v",J" level. Power broadening would have the effect of increasing this fraction. $N_{v'J'}$ is determined from the measured amount of pump power inserted into the v'J' ← v"J" transition within the mode volume. If 1 Watt of 514.5 nm power (2.6×10^{18} photons/sec) is inserted into 0.4 cm³ (contains 86% of TEM_{00} power) and the total dissipation rate out of the v',J' level [Eq. (2)] is 1.5×10^7 Hz, then

$$N_{v'J'} = \frac{(P/h\nu)}{V\tau} = 4.3 \times 10^{11} \text{ molecules.cm}^{-3}. \tag{7}$$

Since the pump power is evenly divided between the (43,0) P(13) and R(15) transitions, $N_{43,16}$ is actually 2.1×10^{11} cm^{-3} compared to

$$N_{v''=0,J''=15} = 5 \times 10^{13} \text{ cm}^{-3}$$

Thus

$$\Delta\nu_{FWHM}^{POWER} \leq \Delta\nu_{FWHM}^{D} \frac{N_{v'J'}}{N_{v''J''}^{0}} = 2.0 \times 10^{6} \text{ Hz}. \tag{8}$$

To summarize

$$2.0 \times 10^{6} \text{ HZ} \leq \Delta\nu^{POWER} \leq 4.0 \times 10^{7} \text{ Hz} < \Delta\nu^{D} = 4.7 \times 10^{8} \text{ Hz}.$$

The looseness of the bounds on $\Delta\nu^{POWER}$ implies that optical pumping causes $N_{v''J''}$ to be significantly smaller than $N_{v''J''}^{0}$. The fact that at 1 Watt of pump power $\mu E/h$ is eight times larger than $\Delta\nu^{H}$ means that the gain lineshape will exhibit structure of the type discussed by Javan [17]. An additional complication, discussed by Schlossberg and Javan [18], arises when the hyperfine structure of the I$_2$ transitions is considered.

2.1.2 Gain Calculation

In estimating the gain of cw optically pumped I$_2$ laser transitions, it is necessary to take into account the fact that each v',J' → v",J" transition is split into either 15 or 21 hyperfine components. The overall width of the hyperfine structure is about 1 GHz, twice as large as the Doppler width. Most of the hyperfine components do not overlap within their homogeneous width. A monochromatic pump laser can be tuned to excite more than half of the hyperfine components, but each is excited via a different velocity group. This means that, for a given v',J' → v",J' laser transition, not all v',J',F' sublevels will be able to contribute to gain. This hyperfine dilution effect is important for weak, narrow bandwidth cw pumping, but is power broadened into insignificance by pulsed pumping.

The stimulated emission cross-section at the peak of a homogeneously broadened transition is

$$\sigma^{0} = 2.6 \times 10^{-19} |R_e|^2 \frac{\nu}{\Delta\nu_{FWHM}^{H}} S_{J'J''} q_{v'v''} \tag{9}$$

where

$$|R_e|^2 = 1(\text{Debye})^2 \text{ at r-centroid } 3.28\text{Å } [16]$$
$$\nu_{43,56} = 1.0 \times 10^{4} \text{ cm}^{-1}$$
$$\Delta\nu_{FWHM}^{H} = 1.6 \times 10^{-4} \text{ cm}^{-1}$$

$S = 0.5$

$q_{43,56} = 4.6 \times 10^{-3}$ [16].

Thus

$\sigma^0 = 3.7 \times 10^{-14} \text{ cm}^2$.

For 1 Watt of 514.5 nm pump inserted and divided equally between the (43,0) P(13) and R(15) lines,

$$N_{v'J'F'} \cong \frac{N_{v'J'}}{21} = 1.0 \times 10^{10} \text{ molecules/cm}^3. \tag{10}$$

Since the active gain length is 50 cm,

$$\sigma^0 N\ell = 1.9 \times 10^{-2} \tag{11}$$

and a single pass gain of 2% is predicted for each of the four (43,56) R(11), P(13), R(15), P(17) lines.

Gain of the strongest (last to be extinguished) laser line in the 1.0 and 1.3μ regions, P(17) of (43,56) and (43,83) respectively, is measured by inserting a quartz window into the I_2 cavity at Brewster's angle and tilting it until the laser is extinguished. Single pass gain is the sum of known output coupling losses and the loss due to the inserted plate calculated from the Fresnel relation. Losses due to scattering and absorption by optics are not known but may be estimated from the observed threshold inserted pump power. The largest observed single pass gain (1 Watt inserted power) and lowest threshold for each transition are

```
(43,56)    3%    110 mW
(43,83)   20%      4 mW.
```

The predicted gain for (43,56) is smaller than the observed value because the hyperfine dilution factor of 21 is probably too large.

It is not possible to predict gain for the (43,83) transition because the electronic transition moment is unknown for the rather large r-centroid, $R_{43,83} = 4.26$ Å, of this transition [16]. However, it is possible to use the observed gain to estimate the transition moment

$$\frac{g_{43,83}}{g_{43,56}} = \frac{|Re(R_{43,83})|^2}{|Re(R_{43,56})|^2} \frac{q_{43,83}}{q_{43,56}} \frac{\nu_{43,83}}{\nu_{43,56}} \tag{12}$$

where

$q_{43,83} = 0.45$ [16].

Thus

$|Re(4.26\text{Å})| = 0.34$ Debye.

2.1.3 Saturation Flux Calculation

If one ignores the effect of the pump radiation field on the gain bandwidth, it is possible to estimate the level of circulating power in the
I_2 gain medium at which the stimulated emission rate for a given
$v'J' \rightarrow v''J''$ laser transition is equal to the loss rate from $v'j'$ due to
spontaneous emission, predissociation, and collisions. Setting

$$\frac{\mu E_{saturation}}{h} = \frac{1}{\pi\tau} = 4.8 \times 10^6 \text{ Hz} \tag{13}$$

1.0µ   1.3µ

$\mu = 4.8 \times 10^{-2}$ Debye $\mu = 1.4 \times 10^{-1}$ Debye

$E_S = 200$ V/cm $E_S = 70$ V/cm

$\omega_0 = 3.9 \times 10^{-2}$ cm^2 $\omega_0 = 4.5 \times 10^{-2}$ cm^2

$P_S = 130$ mW $P_S = 20$ mW.

The circulating power per laser line required for saturation is quite
modest. At 600 mW inserted pump power, circulating power of 5 Watts
divided among perhaps 20 significant laser transitions near 1.3µ is
obtained. It is therefore not surprising that the side fluorescence
from the gain cell increases in intensity by 25% when the rear mirror
is blocked.

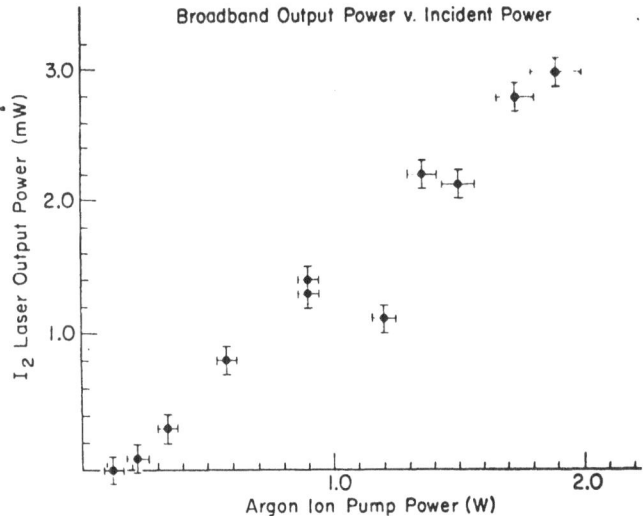

Fig. 4   Multi-line output power from the I_2 laser vs. incident 514.5 nm
power. Power is divided among (v' = 43, v" = 42-56) transitions between
0.8-1.0µ. Note that incident rather than inserted pump power is plotted
and that output coupling and pump polarization were not optimized.

2.2 Problems and Future Experiments

Although the gain and threshold behaviors of the cw I_2 laser are adequately understood, the factors limiting its conversion efficiency are not. Figure 4 shows output power near 1.0μ vs. incident 514.5 pump power. The photon conversion efficiency is only 0.3% but there is no sign of output power saturation. Why is the conversion efficiency so small when the circulating power, 0.3W at 2W pump power, should cause the stimulated emission rate to be comparable to the total loss rate from the upper laser level? There appears to be neither a significant population bottleneck in the lower laser level nor any other intracavity absorptive losses which depend on the presence of pump power.

LOSS MECHANISM

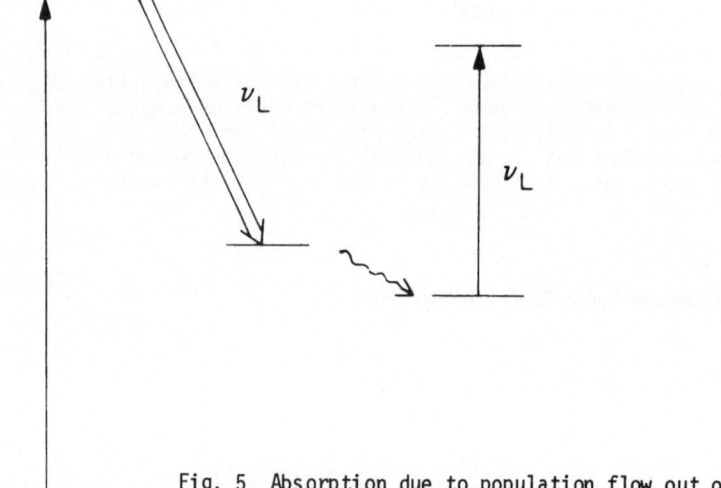

Fig. 5 Absorption due to population flow out of the lower laser level. The I_2 absorption spectrum is extremely dense, especially when highly excited v" levels are populated.

When the incident pump beam was mechanically chopped with a 10^{-7} sec risetime, peak I_2 laser power was observed to be only 10% larger than the average output with the chopper removed.

When the I_2 laser is operated on the higher gain 1.3μ transitions, with optimized output coupling, mode matching, and pump polarization, multi-line photon conversion efficiency of 10% is achieved at 600 mW inserted power. This is a surprisingly high efficiency, considering that a hyperfine dilution factor of 21 would restrict the maximum single line efficiency to $\sim 2\%$. What is the maximum achievable single line efficiency and output power? When does bottlenecking in the lower level set in?

Future optically pumped I_2 laser experiments include the following:

1. An attempt to observe lasing on electronic transitions from BO_u^+ into one of the remaining five unknown gerade states which dissociate into two ground state $I(^2P_{3/2})$ atoms. Mulliken [19] and LeRoy [20] predict that at least one of these states has a van der Waals minimum at about 6 Å. The outer turning point of BO_u^+ v = 43 is near 4.3Å and of v = 62 (excited by the 501.7nm Ar^+ line) near 5.8Å. There should be appreciable vibrational overlap for transitions into the v" = 0 level of a state with r_e = 6Å.

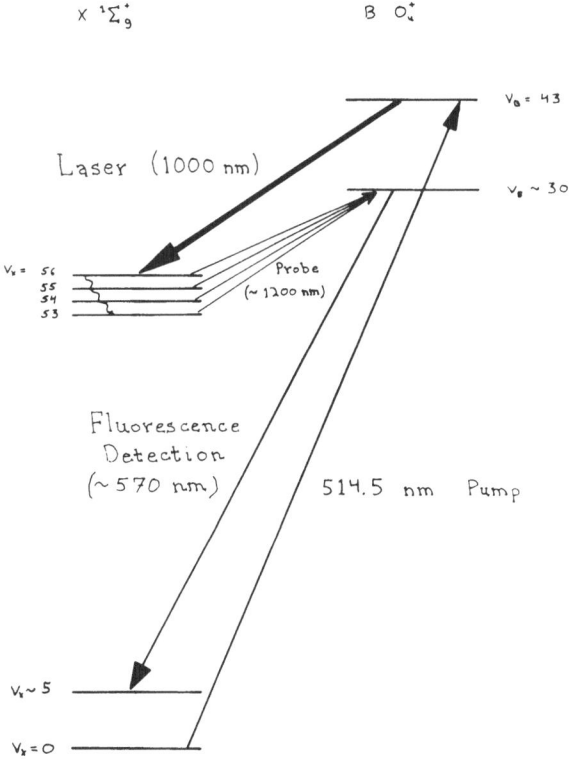

Fig. 6 Pump and probe experiment. I_2 is forced to lase on a transition into a high vibrational level of $X^1\Sigma_g^+$. A tunable probe laser, at wavelengths longer than would be absorbed by unexcited molecules, interrogates I_2 molecules within or near the intracavity mode of the pumped laser. The resultant short wavelength fluorescence is imaged onto the slit of a monochromator tuned to detect a specific (v',v") transition which is not present in the side fluorescence excited by the pump laser. As the probe laser is scanned, populations of various v"+ levels are sampled by excitation of v' ← v"+ transitions. If the pumped laser is Q-switched, the population flux into the v"+ level will be pulsed.

2. An attempt to pump I_2 with the Ar^+ 501.7 nm line which excites the B-X (62,0) R(26) transition. The (62,99) band has a Franck-Condon factor of 0.34 at an R-centroid of 5.8Å [16]. The upper and lower levels of this potential laser transition lie within 100 cm^{-1} of their respective dissociation limits. Observation of such transitions would extend the X state potential beyond the highest previously observed level, v" = 83, well into the chemically interesting long range region. Tellinghuisen[16] also points out that there is a bound free B-X transition from v' = 62 with an exceptionally large 3.5×10^{-2} cm Franck-Condon density about 30 cm^{-1} above the X state dissociation limit. In estimating gain for a bound-free transition, the hyperfine dilution factor becomes unity.

3. With dye laser excitation, it should be possible to obtain a cw I_2 laser oscillating at exactly the frequency of the atomic I laser.

4. Measurement of collisional transfer rates for highly excited vibrational levels of I_2 $X^1\Sigma_g^+$. Figure 6 illustrates the pump and probe experiment by which collisional cross-sections for a variety of (v"+, J"+) ← (v",J") processes are being measured.

Acknowledgements

We wish to thank Professors Robert Byer, Shaoul Ezekiel, Steve Leone, and J.I. Steinfeld for helpful suggestions and advice. We are particularly indebted to Professor Joel Tellinghuisen for providing us with calculated Franck-Condon factors, calling the correct assignments of Ar^+ laser pump transitions to our attention, and enlightening suggestions about power broadening.

References

[1] T.Y. Chang and T.J. Bridges, Opt. Commun. 1, 423 (1970); Appl. Phys. Lett. 17, 249 and 357 (1970).

[2] R.L. Byer, R.L. Herbst, H. Kildal, M.D. Levenson, Appl. Phys. Lett. 20, 463 (1972).

[3] M.A. Henesian, R.L. Herbst, and R.L. Byer, J. Appl. Phys. 47, 1515 (1976).

[4] H. Itoh, H. Uchiki, M. Matsuoko, Opt. Commun. 18, 271 (1976).

[5] F.J. Wodarczyk and H.R. Schlossberg, "An Optically Pumped Molecular Bromine Laser", J. Chem. Phys. 00, 0000 (1977).

[6] S.R. Leone and K.G. Kosnik, Appl. Phys. Lett. 30, 346 (1977).

[7] J.B. Koffend and R.W. Field, "CW Optically Pumped Molecular Iodine Laser", J. Appl. Phys. 00, 0000 (1977).

[8] R. Wellegehausen, S. Shahdin, D. Friede, and H. Welling, Appl. Phys. 13, 97 (1977).

[9] D.R. Guyer and S.R. Leone, 5th Conference on Chemical and Molecualr Lasers, St. Louis, Mo. (1977).

[10] J. Dallarosa, R.E. Drullinger, and J.L. Hall, private communication.

[11] E.J. Schimitschek, J.E. Celto, and J.A. Trias, Conference on Laser Engineering and Applications, Washington, D.C. (1977).

[12] B. Hartmann, B. Kleman, and O. Steinvall, Opt. Commun. 21, 33 (1977).

[13] G.A. Capelle and H.P. Broida, J. Chem. Phys. 58, 4212 (1973).

[14] R.B. Kurzel and J.I. Steinfeld, J. Chem. Phys. 53, 3923 (1970).

[15] R.B. Kurzel, J.I. Steinfeld, D.A. Hatzenbuhler and G.E. Leroi J. Chem. Phys. 55, 4822 (1971).

[16] J. Tellinghuisen, "Intensity Factors for the I_2 B-X Band System", J. Quant. Spectry. and Rad. Transf. 00, 0000 (1977).

[17] A. Javan, Phys. Rev. 107, 1579 (1957).

[18] H.R. Schlossberg and A. Javan, Phys. Rev. 150, 267 (1966).

[19] Robert S. Mulliken, J. Chem. Phys. 55, 288 (1971).

[20] R.J. LeRoy, J. Chem. Phys. 52, 2678 (1970).

Ultraviolet Laser Transitions in I_2 and Br_2

J. Tellinghuisen

Department of Chemistry, Vanderbilt University
Nashville, TN 37235, USA

Introduction

The ultraviolet emission spectra of the halogens have been the subject of
numerous investigations in the last sixty years. (See [1] and [2] and the
references cited therein.) The spectrum of I_2 excited in the presence of
foreign gases has proved particularly challenging to spectroscopists. Some
nine years ago Wieland and I began an isotope shift study of this spectrum,
which very early corroborated the existing analyses of two of the weaker
band systems in the spectrum -- F → X (~2600 A) and E → B (~4300 A). In
the case of the third and most prominent system near 3400 A, the results
indicated that all previous analyses were incorrect. However, because this
system is quite complicated in certain respects, the analysis has not yet
been completed.

A little over two years ago the I_2 3400-A emission system came into pro-
minence of a practical nature as several groups predicted [3] and observed
[4-6] lasing on this transition. Subsequently similar experiments with Br_2
yielded lasing on the so-called E → B transition near 2900 A [7-9]. Strong
emission has also been reported for Cl_2 near 2600 A [10], and very recently
lasing has been reported for F_2 near 1600 A [11].

The similarities in the observed behaviors of the I_2 and Br_2 laser transitions suggest that they may have the same electronic designations in the two
molecules. Yet our analysis of the I_2 3400 system eliminated the B state
from involvement in this spectrum (at least at the long-wavelength end where
the strongest bands occur). In this light the E → B designation for the Br_2
2900 system comes into question. Noting that previous work on this spectrum
had been done with natural bromine (which is roughly 50:50 $^{79}Br:^{81}Br$), I
decided to reexamine the spectrum using a single Br isotope. The preliminary phase of this study has now been completed, and a brief account has
been published [12]. The main result of this effort is the verification that
the E → B designation is wrong for Br_2. Instead, for both molecules this
system is very likely ascribable to the 1432, 2g ($^3\Pi$) →2431, 2u ($^3\Pi$) transition, which I am abbreviating as D' → A'. (The 4-digit prefixes are Mulliken's
designation for the simplified MO descriptions of these states. The Hund's
case c designations are emphasized, since these states probably approximate
this coupling in these heavy molecules; the probable case a designations are
given in parentheses.) In the remainder of this paper I will discuss certain

aspects of these transitions which are particularly relevant to the laser developmental work.

Discussion

The results of the vibrational analyses for both molecules are summarized in Table 1. The vibrational expansion coefficients bear the usual significance,

$$\nu_i = \Delta T_e + \sum_{j=1}^{m} c_j' \left[\rho(v_i' + 1/2)\right]^j - \sum_{k=1}^{n} c_k'' \left[\rho(v_i'' + 1/2)\right]^k, \quad (1)$$

where ν_i is the measured band head and ΔT_e, $\{c_j'\}$, and $\{c_k''\}$ are the constants given in the table. The quantity ρ is the usual isotopic factor, $\rho = (\mu/\mu')^{1/2}$, and is referenced to $^{127,127}I_2$ and $^{79,79}Br_2$ (i.e., $\rho = 1$ for these isotopic species). The constants were obtained by least-squares fits of all assigned bands for the several isotopic species of each molecule. In this procedure the v'' numbering was determined by varying a trial numbering until minimum variance was achieved. For I_2 the assigned bands spanned v' levels 0 - 8 and v'' levels 5 - 23 (λ = 3275 - 3450 A). For Br_2 the assignments covered v' = 0 - 5, v'' = 5 - 15, and λ = 2800 - 2950 A. Although only red-degraded band <u>heads</u> were included in the fits, simple calculations show that the origins lie within 0.2 cm^{-1} of the heads for these bands. For both molecules the best fits were obtained with m = 2 and n = 4.

Table 1 Spectroscopic constants for ultraviolet emission bands in I_2 and Br_2. Standard errors (in parentheses) are in terms of the last significant digit. All quantities are in units cm^{-1}.

	$^{127,127}I_2$	$^{79,79}Br_2$
ΔT_e	30341.0 (16)	35677.1 (74)
c_1'	103.97 (5)	150.79 (10)
c_2'	-0.205 (6)	-0.356 (19)
c_1''	106.29 (54)	152.82 (306)
c_2''	-0.856 (63)	-0.471 (460)
c_3''	-3.016 (310) x 10^{-2}	-0.1058 (299)
c_4''	5.52 (55) x 10^{-4}	1.68 (71) x 10^{-3}
σ	0.28	0.23
D_e''	2350 (100)	2830 (150)
T_e''	10200 (100)	13230 (150)
T_e'	40540 (100)	48910 (150)

The absolute energies T_e', T_e'', and D_e'' are imprecise because (a) none of the states involved in these transitions has been identified in transitions to other known states of these molecules, and (b) a lengthy extrapolation is required to estimate D_e''. According to the present assignment the lower state (A') of both transitions is the lowest excited state in both molecules and dissociates to two ground state ($^2P_{3/2}$) atoms. At high pressures the excited state (D') seems to have a sort of preferred status among the ion-pair states, in that the excitation energy seems to pool in this state and remain there until radiative decay occurs. It is tempting to suggest that D' is the lowest of the ion-pair states in the halogen molecules; however further work will be required to verify this point.

Figure 1 shows the potential diagram for the Br_2 D' → A' transition and indicates the main bands which appear in the laser spectrum. The diagram

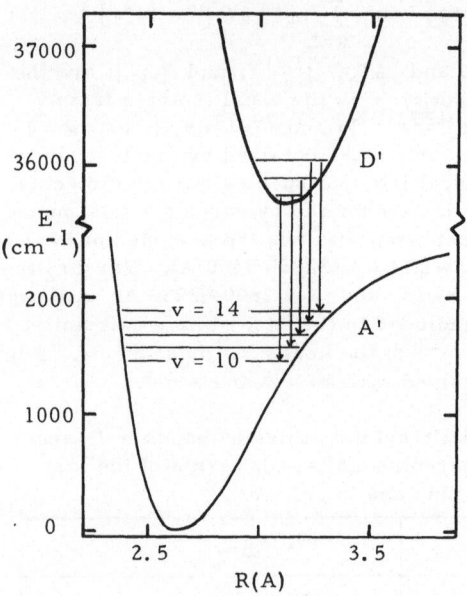

Fig. 1 Potential diagram illustrating main transitions occurring in the Br_2 UV laser. Energies are relative to the minimum of the A' curve.

for I_2 is similar, with lasing occurring mainly on the 0-12, 2-15, 3-17, 1-14, and 0-13 bands. In both cases lasing must actually occur on the densely overlapped rotational lines of these and other nearby bands, and in Br_2, on the bands of three different isotopic molecules. It seems likely that cavity tuning could be used to extend the wavelength ranges of both lasers.

Although the radiative lifetime has not yet been measured for the D' state of either of these molecules, a reasonable guess is $\tau \approx 20$ ns in both cases. These transitions are inherently strong because of their charge-transfer character. In I_2 the D → X (1800 - 3300 A) and E → B (3500 - 4400 A) transitions are of the charge-transfer type, and the D- and E-state lifetimes have been measured to be 16 ns [13] and 27 ns [14] , respectively. Franck-Condon factors for the strong bands in the laser spectrum are typically of magnitude 0.2 - 0.3. These figures can be used to estimate gain coefficients.

The I_2 and Br_2 UV lasers have so far failed to demonstrate the high power and efficiency one might predict from studies of spontaneous emission yields [3]. The reason for such meager performance has not been determined, but a naive suggestion is that these systems suffer from loss processes in which the laser radiation is absorbed by the molecules in their lasing state (D'). Clearly additional work will be needed to assess the roles of these and other loss processes and possibly improve the performance of these lasers.

References

1. K. Wieland, J. B. Tellinghuisen, A. Nobs: J. Mol. Spectrosc. 41, 69 (1972)
2. R. S. Mulliken: J. Chem. Phys. 55, 288 (1971)
3. M. V. McCusker, R. M. Hill, D. L. Huestis, D. C. Lorents, R. A. Gutcheck, H. H. Nakano: Appl. Phys. Lett. 27, 363 (1975)
4. R. S. Bradford, Jr., E. R. Ault, M. L. Bhaumik: Appl. Phys. Lett. 27, 546 (1975)
5. J. J. Ewing, C. A. Brau: Appl. Phys. Lett. 27, 557 (1975)
6. A. K. Hays, J. M. Hoffman, G. C. Tisone: Chem. Phys. Lett. 39, 353 (1976)
7. D. C. Lorents, private communication
8. J. R. Murray, J. C. Swingle, C. E. Turner: Appl. Phys. Lett. 28, 530 (1976)
9. J. J. Ewing, J. H. Jacob, J. A. Mangano, H. A. Brown: Appl. Phys. Lett. 28, 656 (1976)
10. C. H. Chen, M. G. Payne: Appl. Phys. Lett. 28, 219 (1976)
11. A. K. Hays, private communication
12. J. Tellinghuisen: Chem. Phys. Lett. (in press)
13. A. B. Callear, P. Erman, J. Kurepa: Chem. Phys. Lett. 44, 599 (1976)
14. D. L. Rousseau: J. Mol. Spectrosc. 58, 481 (1975)

Gas Phase and Other New Dye Laser Developments

F.P. Schäfer

Abteilung Laserphysik, Max-Planck-Institut für biophysikalische Chemie
3400 Göttingen, FRG

In the last few years dye lasers have become very important
tools for spectroscopy, and very recently interesting photochemical applications have become known and will be of increasing importance in the future, laser isotope separation being
only one of these, albeit the most publicized. If these applications are to have any economic importance at all, the efficiency of the dye lasers used in these applications should be
as high as possible, certainly higher than the half percent or
so, which is the state of the art in today's dye lasers.

If one now asks what is the reason for this low efficiency,
one finds that the optical excitation is to be blamed. Pumping
a dye laser with the monochromatic emission of another laser,
e. g. a nitrogen laser or an argon ion laser, the transforma-
tion of the pumping light flux into dye laser emission can
easily have an efficiency of 35 %, so one sees that the intrinsic efficiency of dye lasers is very high indeed. But in the
examples cited, the efficiency of the pump laser is very low,
so that the overall efficiency is deplorably bad.

This is much better for the case of incoherent pumping,
using xenon flashlamps, which can transform about 30 to 50 % of
the electrical input energy into light output covering the
whole range from the near ultraviolet over the visible to the
near infrared. Unfortunately, only a small fraction of this
broadband radiation can be absorbed by the dye molecules, so
that again the overall efficiency can be at most one percent.
It is not necessary to go into the details of the various
possibilities to improve this situation and only mention in
passing that there are good prospects for synthesizing dyes
with a broadband absorption spectrum, that can make better use
of the white light of the flashlamps, as well as on the other
hand to develop new high intensity light sources with an output
that is spectrally matched to the absorption spectra of the
dyes now in use.

But these improvements will be of limited value only, and
one immediately comes to the question whether the optical excitation could possibly be circumvented altogether by a direct
electrical excitation of the dye molecules. One indeed has some
scarce information about such processes. In liquid solution
there is the so-called electro-chemiluminescence, which means
that by applying an electric field to a dye solution, the neutral dye molecules form ions at the electrodes, which drift
together in the electric field and emit fluorescence on recombination. There were also some early investigations with low
energy and high frequency discharges in some buffer gases containing some dye vapors, which showed a weak fluorescence coming

from the dye molecules. There are many intrinsic limitations of the electro-chemiluminescence process which, however, will not be discussed here. In the vapor phase processes such limitations are not immediately obvious, and this is the reason, why several groups have taken up the development of vapor phase dye lasers with the ultimate goal of an efficient direct discharge laser.

But there is still another good reason for the development of a vapor phase dye laser and that is connected with the wavelength of dye lasers. The shortest wavelength of dye lasers should be that obtained with substances containing just one double bond, like ethylené and its derivatives, which absorb around 180 nm, so that with the usual Stokes shift one could expect fluorescence around 200 nm. At wavelengths below 350 to 300 nm many solvents start to absorb strongly, and at 200 nm there are very few solvents available and then only few dyes are soluble in these special solvents. So for the achievement of shorter wavelength dye laser emission below the present limit of 322 nm one would also better try to use dyes in the vapor phase, especially since the absorption is usually blue-shifted going from any solvent over to the gas phase as we shall see immediately.

There are additional advantages of a less fundamental nature, like the avoidance of schlieren effects created in dye solutions with non-uniform illumination, because of the high temperature dependence of the refractive index of the solvents, but clearly the two main incentives to study vapor phase lasers are the two prospects of higher efficiency and shorter wavelengths in dye lasers.

A detailed account of the published work on vapor phase dye lasers is to be found in [1]. Recent work on electron beam excitation of dye vapors was discussed and is in press [2].

In conclusion, it was found that vapour phase dye lasers have demonstrated their potential with optical pumping and progress with electron pumping seems quite encouraging, but it is probably still a long way towards a direct discharge-excited vapor phase dye laser.

Finally, the development of a 100 W average power dye solution laser pumped with flashlamps in a novel design was discussed and performance data given. A paper on this laser is in press [3]. It was concluded that this design could easily scaled to an average output power of over 1 kW.

References

1. F. P. Schäfer (Ed.): *Dye Lasers*, Topics in Applied Physics, Vol. 1, 2nd Edition, (Springer-Verlag Berlin, Heidelberg, New York 1977) p. 266
2. G. Marowsky, R. Cordray, F. K. Tittel, W. L. Wilson, and J. W. Keto: J. Chem. Phys. (in press)
3. J. Jethwa, F. P. Schäfer, J. Jasny: IEEE J. Quant. Electr., Febr. 1978 (in press)

Injection-Locked, Unstable Resonator Dye Laser

I.J. Bigio*

University of California, Los Alamos Scientific Laboratory
Los Alamos, NM 87545, USA

In pulsed dye lasers intended for photochemistry (for example) the requirements of large pulse energies (> 100 mJ) are inconsistent with those of
narrow spectral linewidth and low divergence. With commonly used stable
resonators, a low Fresnel-number cavity is usually required for single,
transverse-mode operation, yielding small mode volume and, therfore, low
pulse energy. The larger mode volumes desired for energy extraction in
flashlamps-pumped lasers can only be achieved, with single-mode operation in
a stable resonator, by the use of intracavity lenses and spatial filters[1]
which reduce efficiency. The intracavity dispersive elements used for line
narrowing also greatly reduce efficiency, and often place peak-power limitations because of low damage thresholds of some of the elements.

The use of an unstable resonator [2] can combine the advantages of large
mode volume with single, lowest-order mode operation, yielding a near
diffraction-limited beam. The positive-branch confocal resonator is particularly useful in that it gives a collimated output beam and has no internal
focal points. (See Fig. 1).

CONFOCAL UNSTABLE RESONATOR

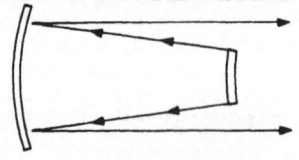

POSITIVE BRANCH

Fig. 1 The most convenient unstable
resonator configuration

However, unstable resonators do not lend themselves readily to conventional tuning and line-narrowing techniques. Those methods which can be employed in unstable resonators greatly complicate the otherwise simple cavity
and, again, place limitations on peak pulse powers. However, injection locking can be used to force the laser to run at the wavelength and bandwidth of
a low power injected beam, while eliminating the need for any dispersive
elements.

Injection locking of pulsed dye lasers with a cw, narrow-band laser has
been analyzed by GANIEL, et al. [3], and also performed experimentally [1].

*
This work was done at the Weizmann Institute of Science, Department of
Electronics, while the author was supported by a Fullbright Senior Lectureship.

While achieving the desired bandwidth of under 30 MHz (equal to the band-width of the injected beam) they were limited to pulse energies of 10-50 mJ because of the complicated stable, ring-resonator employed. This was neces-sary for single-mode operation. Since the injected light must compete with build-up from noise in *all* transverse modes, the injected power required to reliably lock is proportional to the number of transverse modes, m, which oscillate in the laser cavity. Thus,

$$P_{inj} \propto m = CF^2$$

where C is a constant and F is the Fresnel number of the cavity. Therefore, single-mode operation minimizes required injection power in addition to optimizing beam divergence. The injection power required for complete locking will also depend on the gain difference between the free-running peak and the locked wavelength, as well as on the pulse duration and total cavity losses.

Injection locking of unstable resonators was first suggested by SEIGMAN [2] and described experimentally by BUCZEK, et al. [4], who used low power CO_2 lasers with a resulting power ratio, $P_{out}/P_{inj} \sim 100$ (possibly limited by lack of longitudinal-mode matching, among other causes). BLIT, GANIEL, and TREVES [1], with their stable resonator, achieved power ratios around 3×10^7 by carefully matching the longitudinal modes of the cw and pulsed laser cavities. The major improvements here over the latter work are due to the use of the unstable resonator for the coaxial flashlamp-pumped dye laser. This gives near diffraction-limited pulse energies of ≥ 1 joule (with Rhodamine 6G) for a stored energy in the capacitor of 140 joules, yielding an overall efficiency of 0.7%. The pulse duration is about 300 ns; thus, with an injected cw power of a few milliwatts we get $P_{out}/P_{inj} \sim 10^9$.

The experimental arrangement is shown in Figure 2. The cw beam is

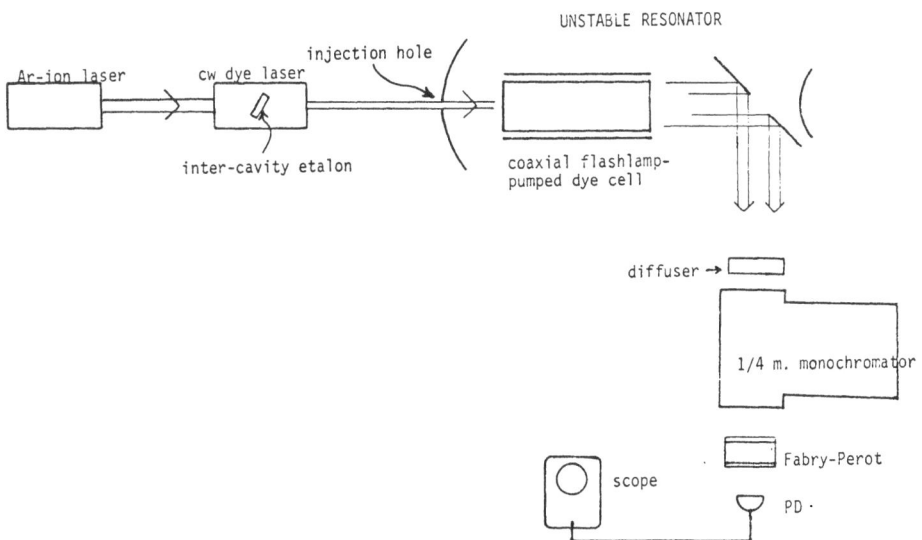

Fig. 2 Block diagram of the experimental setup.

injected through a 1 mm diameter hole in the concave mirror, and is coupled to the rest of the cavity volume by the inherent geometric magnification of the cavity, in addition to diffraction-caused expansion. The output coupling is by a 45° mirror with an elliptical hole. The bore of the dye cell is 8 mm, and losses through the injection hole are small compared with the effective output coupling traction of 70%. One side benefit of this arrangement is that the often tedious chore of aligning a small-bore unstable resonator is simplified by use of the on-axis cw beam. As was shown in [1], reliability of the injection locking depended strongly on the longitudinal-mode matching, and simplified methods for achieving this are being investigated.

Thus, it is possible to *simultaneously* obtain large pulse energy, low divergence, and very narrow linewidth in a flashlamp-pumped dye laser. In addition to the obvious applications in photochemistry, the tunability feature and single-mode operation also make it useful for contour holography [5].

The author would like to express appreciation to Yosef Kedmi, who assisted in many of the experiments, and especially to Prof. David Treves for his hospitality and for many stimulating converstions.

References

1. S. Blit, U. Ganiel, and D. Treves, Appl. Phys. 12, 64 (1977).

2. A. E. Siegman, Appl. Opt. 13, p. 353 (1974).

3. U. Ganiel, A. Hardy, and D. Treves, IEEE J. Quant. Elec. QE-12, 704 (1976).

4. C. J. Buczek, R. J. Freiberg, and M. L. Skolnick, Proc. of the IEEE, 61, 1411 (1973).

5. As suggested by Prof. A. A. Friesem, Weismann Inst. of Science, Rehovot, Israel (private communication).

Short Wavelength Multiline Performance of CW Ion Lasers

H.R. Lüthi and W. Seelig

Institute of Applied Physics, University of Berne
Berne, Switzerland

Abstract

High power cw laser oscillation on 7 ArIII and 3 KrIII lines
below 350 nm is reported. The active medium used is a highly
ionized low pressure dc discharge (length: 1.7 m, radius: 6mm)
from which 75 Watts in the combined 351.1 nm and 363.8 nm ArIII
lines can be extracted. For every line under investigation the
output increases without saturation up to the available discharge current i = 480 A. The maximum power values are: 17 W
(ArIII 333-335 nm); 3.8 W (ArIII 300-305 nm) and 0.4 W (ArIII
275 nm).

Sources of coherent ultraviolet radiation, emitting more
than 1 W cw power, are attractive for many applications such
as spectroscopy, dye laser pumping, photochemistry and isotope
separation. At present ion lasers are the most powerful cw
sources in this region. More than one hundred ultraviolet ion
laser lines have been reported [1,2] and recently 60 Watt cw
emission was realised with an ArIII- (351.1 and 363.8 nm) -
laser [3]. From the other UV laser lines substantially less
cw power has been generated and below a wavelength λ = 325 nm
no cw power measurements have been reported [4]. Therefore we
have performed a series of experiments to increase both, the
power and the efficiency of different ion laser transitions
down to λ = 275 nm in Ar- and Kr-arc-discharges.

The wall stabilized arc-discharge for UV laser generation is
similar to that used for an ArII-laser in the 200 Watt output
range [5,6]. The discharge parameters for high power operation
of UV laser transitions in doubly ionized Argon and Krypton
are electron temperatures in the range $kT_e \simeq$ 4...4.5 eV, electron densities $n_e \simeq 1.5 \cdot 10^{14}$ cm^{-3} and a degree of ionization
$n_e/N \gtrsim 0.5$ (N: neutral gas density). The discharge tube which
confines the highly ionized low density plasma has a length of
1.7 m and is composed of water cooled anodized aluminium seg-
ments [5] having a 12 mm discharge bore and a length of 25 mm.
At the maximum available current of the power supply i_{max} =
480 A, the electrical power input is 1.2 kW per cubic centimeter of the discharge volume.

While the output power of the visible ArII-laser transitions
is limited due to resonance trapping effects [7], the performance of ArIII and KrIII-UV-ion lasers is limited by discharge
instabilities and can be improved by an increase of the stable

discharge range [5,8] . The stability limit of ion laser gas discharges is determined by neutral gas clean up, setting an upper limit to n_e/N [9] . Other effects which produce instabilities are either negligible in the range under investigation (self magnetic field) or can be avoided by a careful choice of the geometry of the transition regions between the arc column and the electrode vessels and by the use of a bypass tube which removes the influence of pressure pumping effects. After optimisation of the geometry, we found that the stability limit can be raised with the aid of an axial magnetic field in the 20 Gauss range [8] . In this manner the discharge under investigation could be operated in the stable region up to the maximum available current of the power supply.

This discharge technique has been used up to now for the excitation of 9 Argon and 5 Krypton high power UV laser transitions. The optical resonator (length: 3.4 m) consisted of two dielectrically coated concave quartz mirrors both having a 6 m radius of curvature. Each wavelength group was investigated with a pair of mirrors, one having a reflection coefficient R ≈ 99.9% and the other one in the region 97.5% < R < 99% (Table 1).

Table 1 Measured UV laser power at i_{max} = 480 A. Discharge diameter 12 mm (discharge length: 1.7 m, fill pressure: 1.25 torr for Ar, 1.3 torr for Kr)

Ion	Wavelengths nm	Power Watts	Output Reflector %	Relative Contribution %
ArIII	363.7 351.1	61	98	50 50
ArIII	335.8 334.4 333.6	17	97.5	25 45 30
ArIII	305.4 302.4 300.2	3.8	99	5 40 55
ArIII	275.4	0.4	98.5	100
KrIII	356.4 350.7	19	98	30 70
KrIII	337.4 323.9 312.4	4.5	97.5	25 70 5

The high reflection coefficients of the outcoupling mirror coatings are necessary because of the low gain coefficients which are below 0.1% per cm active discharge length [6].

The figures 1 and 2 show the measured current dependence of the output power per unit length of the discharge. The fill pressure is optimised for each current position. The power of all investigated lines is rising up to i_{max} = 480 A corresponding to a jR product of 255 A/cm (j: current density). The importance of the stabilisation is strongly demonstrated by the

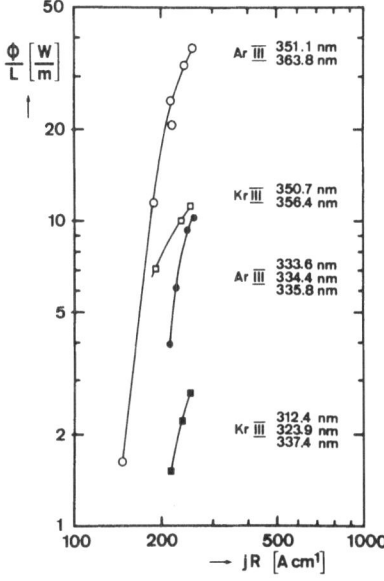

Fig. 1 Measured jR-dependence of the UV laser output power Φ per unit length of the discharge. (λ > 310 nm, discharge diameter: 12 mm, fill pressure optimised for each current position)

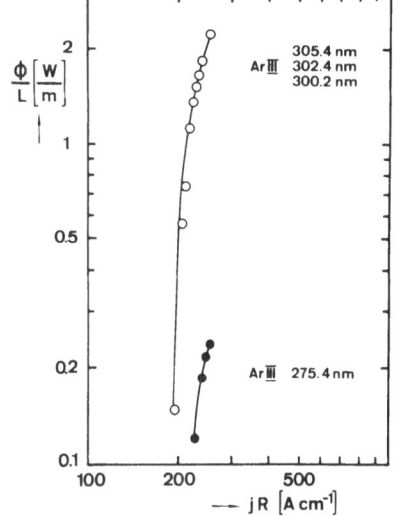

Fig. 2 Measured jR-dependence of the UV laser output power Φ per unit length of the discharge. (λ < 310 nm, discharge diameter: 12 mm, fill pressure optimised for each current position)

results obtained with the short wavelength transitions $\lambda <$ 305 nm which have their thresholds around jR $\simeq$ 200 A/cm.

Table 1 gives the maximum power values of each line group. The highest power was 61 Watt from the combined 351.1 nm and 363.8 nm Argon lines. For this power level the efficiency amounts to 0.03% which is 3 times higher than previously reported [10] . The observed power from the three Ar lines around 334 nm and from the 337.4 nm KrIII line is more than one order of magnitude higher than reported previously [11] . cw operation of the ArIII- and KrIII-lines 323.9 nm, 312.4 nm, 300.2 nm and 275.4 nm has already been reported by MARLING without power data, however [4,12] . The ArIII-lines 305.4 nm and 302.4 nm have not been observed in cw operation before.

A further output increase for the 14 UV laser lines under investigation could be limited by saturation effects in the laser mechanism [7] or by discharge instabilities [6,9] . However, for most of the cw lines the maximum power obtained in pulsed discharges (0.2 - 0.3 µsec duration) was found to rise with the discharge current density up to values of jR and Φ/L which are both one order of magnitude higher than in our experiments [4] . Therefore the depopulation of the laser levels by electron collisions seems to be negligible for currents up to at least 600 A in our 12 mm diameter arc. Also any bottleneck by radiation trapping [7] should be negligible because the density in the ArIII-groundstate is much smaller than in the ArII-groundstate.

A quantitative prediction of the power limit for the UV ion laser is difficult because the effects of discharge instabilities at currents higher than 480 A are unknown. Therefore we are presently modifying our power supply in order to achieve a value of i_{max} = 600 A. Very recently i_{max} was increased in a first step from 480 A to 500 A. Thereby the output power of the ArIII (351.1 and 363.8 nm) laser was rising up to 75 Watts.

Acknowledgement

The authors are indebted to Prof. H.P. Weber for his permanent support, to Dr. W. Lobsiger (Balzers AG Liechtenstein) for preparation of the dielectric mirror coatings and to J. Steinger for experimental assistance. This work was sponsored by the "Kommission zur Förderung der wissenschaftlichen Forschung des EVD" of Switzerland.

References

1. C.C. Davis and T.A. King, "Gaseous ion lasers", in Advances in Quantum Electronics, Academic Press, London and New York (1975)

2. G.J. Collins, Journal of the Optical Society of America 66, 1100 (1976)

3. H.R. Lüthi, W. Seelig and J. Steinger, submitted for publication to Applied Physics Letters

4. J.B. Marling, IEEE Journal Quantum Electron. QE-11, 822 (1975)

5. G. Schäfer and W. Seelig, Z. angew. Physik 29, 246 (1970)

6. H.R. Lüthi, J. Appl. Phys. 48, 664 (1977)

7. H.R. Lüthi and W. Seelig, Appl. Phys. 6, 261 (1975)

8. H.R. Lüthi, W. Seelig, J. Steinger and W. Lobsiger, IEEE J. Quantum Electron. QE-13, 404 (1977)

9. H.R. Lüthi and W. Seelig, submitted for publication to J. Appl. Phys.

10. J.R. Fendley, Jr. and J.C. O'Grady, "Development construction and demonstration of a 100 watt cw Argon ion laser", United States Army Electronics Command, Research and Development Technical Report ECOM-0246-F (1970)

11. J.R. Fendley, Jr., IEEE Journal Quantum Electron. QE-4, 627 (1968)

12. J. Marling, private communication.

High Power Iodine Atom Laser

K. Hohla

Projektgruppe für Laserforschung der Max-Planck-Gesellschaft
zur Förderung der Wissenschaften e.V.
8046 Garching, FRG

1. Introduction

Thermonuclear fusion by means of laser beams calls for laser systems with extraordinary properties. These can be characterized under the headings: high power, short pulse, good beam quality and high efficiency. The most powerful systems available today are Nd-glass lasers in the range of 1 - 2 Terawatts. And laser systems are under construction with a tenfold increased intensity at an energy level of about 100 kJ. Among the very many laser substances there are only three which fulfil the conditions characteristic for high power lasers. These are the Nd-glass-, the CO_2- and the iodine-laser. Considering the enormous amount of money necessary to build lasers in the 10 to 100 kJ range one has to look critically at the chances for a further development of these systems. Up to now it seems that none of the three presently available lasers is a candidate which can be employed in a laser fusion reactor. But the exploration and development phase for a new laser - the laser X - will need at least 10 years - according to the bad experience with the other systems. So the next generation of high power lasers to be used for scientific breakeven experiments will either be a Nd-glass-, CO_2- or iodine-laser. In this paper the iodine laser will be described in more detail emphasizing its future prospects and its developmental possibilities.

2. Basics of the Iodine Laser

According to a photochemical process discovered by Kaspar and Pimentel [1] in 1964 alkyliodides irridiated with UV-light split off the iodine atom (see Fig.1).

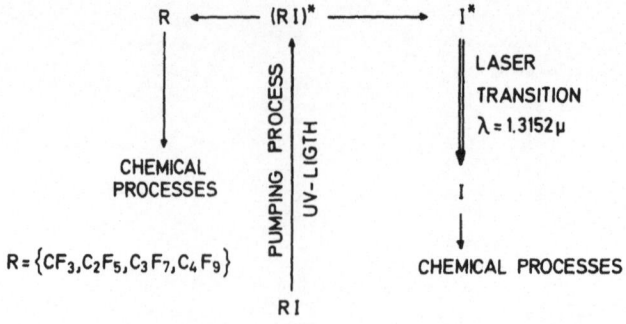

Fig.1 Processes leading to laser action in an iodine laser

The iodine atom is in the first electronically excited state($^2P_{1/2}$ state). The transition to the groundstate $^2P_{3/2}$ is forbidden, the upper state having the lifetime of 170 ms. Unfortunately this lifetime cannot be used completely. The reasons are quenching processes caused by the parent molecules and by the reaction products. For the same reason gas additives which are sometimes necessary can also shorten the lifetime. The transition from the $^2P_{1/2}$ to the $^2P_{3/2}$ state, the laser-transition, shows a more complex structure as can be seen from Fig.2 [2] . The upper state splits in the sublevels F=3 and F=2 under the influence of the magnetic field of the nucleus. The degeneracy of these states is 5 and 7 according to the momentum of the nucleus. The F number and the degeneracy g of the lower state are also given in Fig.2. The spectrum of the 6

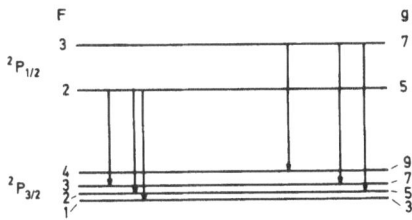

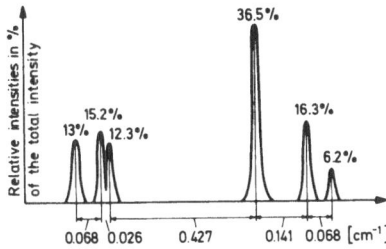

Fig.2 Level diagram of the iodine atom laser (upper part) and relative intensities of the transitions (lower part)

different transitions, according to $\Delta m = 0, \pm 1$, is shown in the lower part together with the relative intensities. The strongest transition starts from the F=3 and ends at the F=4 level. Normally this line oscillates in an iodine laser. The stimulated emission cross section σ being one of the most important laser parameters, depends critically on the pressure. Above 10 Torr laser gas pressure the transition is pressure broadened and can therefore be varied by adding a foreign gas [3] . In Fig.3 the cross section is shown as a function of the gas pressure. As one can see adding CO_2 to the lasergas is nearly twice as efficient as adding Ar. The dashed curves are measurements for Ar made by Aldridge [4] and by Zuev and coworkers [5] . They compare not too bad with the measurements by Fuß [6] . This means that σ can be changed very easily in the iodine laser and this within more than one order of magnitude. This variation of σ can be performed independently on the energy content of the laser. Naturally one has to take care of the deactivation by the foreign gas. In Table 1 a list of several gases with their quenching and broadening characteristics is given [3] . The last three columns show the maximum tolerable pressure at

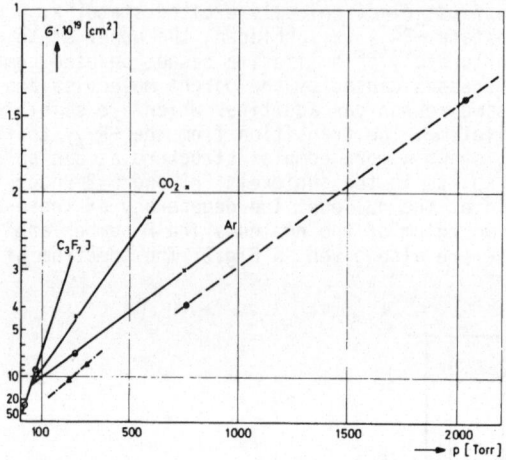

Fig.3 Cross section for stimulated emission as a function of partial pressure of Ar and CO_2

x, ⊙ : W. Fuß and K. Hohla
: V. Zuev et al.
: Aldridge

Table 1 Rate coefficients k_M for collisional deactivation of I*, tolerable pressures p_{10} at which the deactivation reaches 10 % in 10 µsec, and cross-sections σ for p_{10} and for p = 700 Torr foreign gas pressure.

M	$k_M/10^{-16}$, cm^3/sec	reference	p_{10}Torr	$\sigma/10^{-19}$ cm^2 for p_{10}	for 700 Torr	reference
He	0.02	9	170,000		4.6	7
Ar					3.8	7
					4.0	4
					3.5	2
	0.02	9	170,000		4.0	6
N_2	2	9	1,700	1.15	2.8	6
CO	12	9	280	5.7		6
CO_2	1.5	9	2,200	0.65	2.0	6
	4.6		700		1.95	6
SF_6	0.24	9	14,000		2.9	6
CF_2Cl_2	25		135	6.6		6
CF_3I	2.2	6	1,500	0.77	1.64	6
	0.65	8				6
i-C_3F_7I	8.0	8	420	1.5		6
CF_3Br	a		(600)[a]	3.1		6
$(CF_3)_2CO$	a		(350)[a]	2.0		6
I, I*						6

[a]Nonexponential decay is found with CF_3I (see ref. 6).

which the deactivation reaches 10 % within 10 µs, and the cross section for this pressure p sub 10 and for 700 Torr foreign gas.

A numerical analysis of the six transitions [3, 10] shows that the lines overlap at pressures of several hundred Torrs. In Fig.4 the line profiles for three different CO_2 partial pressures are given-for higher pressures than 500 Torr CO_2, we have only one transition which can be considered as homogenously broadened.

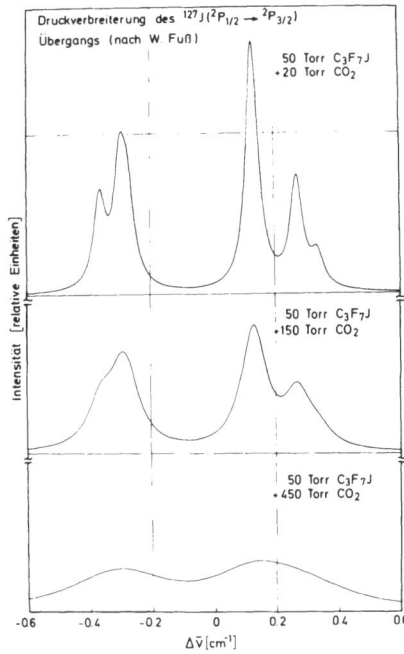

Fig.4 Calculated line profiles of the iodine transition for three different partial pressures of CO_2

At the end of this short summary a few words to the photolysis. As already mentioned it occurs in the UV-range and more exactly, as shown in the Fig.5, between 2500 and 3000 Å. The peak absorption is shifted to longer wave lengths for higher alkyliodides and the bandwidth grows a little bit. C_3F_7I, the most frequently used alkyliodide, has a width (FWHM) of 300 Å and peaks at an absorption cross section of 6×10^{-19} cm^2. Most of the laser are driven with Xenon flash lamps which have an efficiency of about 8 % in this spectral range.

3. Generation of High Power Laser Pulses

All the high power lasers, Nd, CO_2, and iodine, work on the energy storage principle, which is as follows: The required pulse length of approx. 1 ns is several orders of magnitude smaller than the pump time of these lasers. During the pulse length practically no energy is therefore pumped into the laser; instead, it has to be stored in the system before the laser beam passes. This process is illustrated in the upper part of Fig.6. The laser is pumped for approx. 1 to 200 µs depending on the laser medium. The inversion is stored in the laser material and is then available for amplifying the incoming beam.

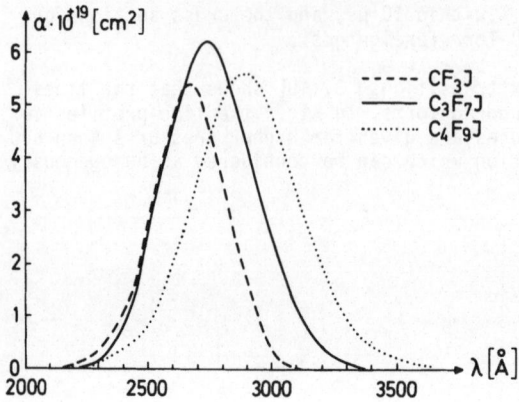

<u>Fig.5</u> UV-absorption spectra of CF_3I, $i-C_3F_7I$ (mostly used) and $t-C_4F_9I$

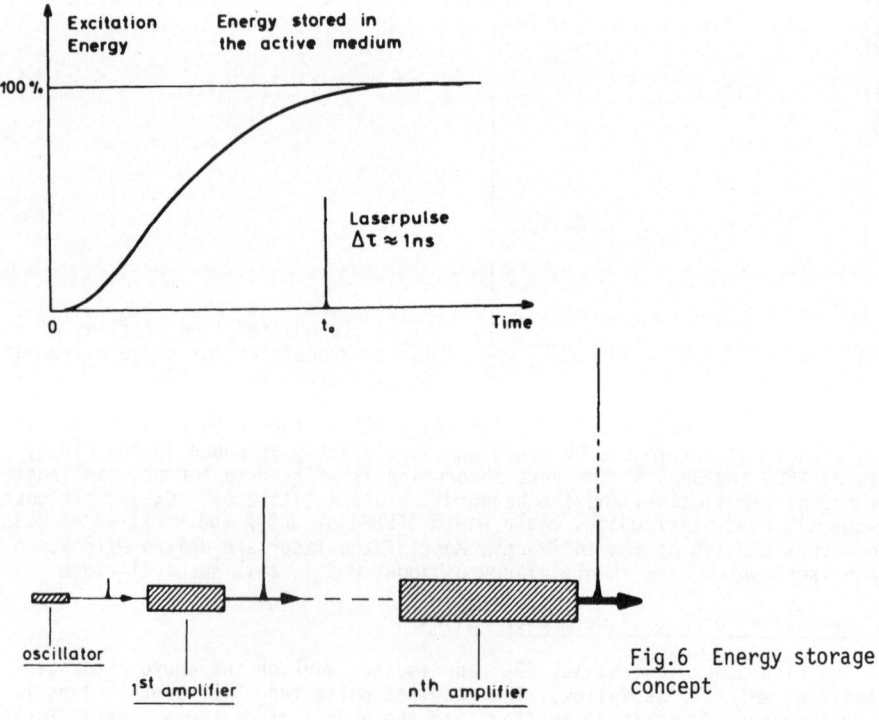

oscillator

1st amplifier

n^{th} amplifier

<u>Fig.6</u> Energy storage
concept

In keeping with this scheme, high-power lasers consist of an oscillator, which
generates the pulse to be amplified, backed up with several amplifiers which
is schematically depicted in Fig.6.

The real problems involved in high power lasers concern

a) the generation of short pulses
b) the pulse propagation of the short pulse in the amplifier
c) scalability and efficiency of the amplifier.

4. Short Pulse Generation

The requirements for pulse duration and pulse shape are set by laser fusion experiments. At this time pulse durations with a FWHM of 100 ps seem to be desirable, the risetime being as steep as possible. These requirements will change in the future. Larger pulse energies will be desired in longer pulse durations. 100 kJ systems should have durations of 0,5 to 1 s. But neverthe- less, first the question will be discussed how 100 ps pulses can be generated what is the state of the art, what possibilities show up. The most prominent method at the moment - because of the lack of useful saturable absorbers - is active modelocking. Acoustooptic and electrooptic methods have been used. In many laboratories like in Sandia [11] or in Manchester [12] or in Garching [13] pulse durations with 400 - 800 ps have been achieved by active mode- locking. A considerable reduction of the pulse duration employing this method can only be expected by using higher pressures of 5 - 10 atm of the foreign gas. A principal disadvantage of this method is the poor reproducibility of these pulses. Among the many other methods to generate shorter pulses one possibility deserves special consideration: the method of free induction decay (FID) [14] . It is based on truncating the pulse by optical breakdown and fil- tering the narrow band frequency components of the original pulse by a narrow band filter.

In Garching the following experiment was performed (Fig.7) [15] : A 3 ns pulse was focussed into an I_2 cell. Termination of the pulse by optical break- down is especially fast if the f-number of the focussing optics is small.

The ultrafast fall time of the pulse gives rise to a broad spectrum. In a hot iodine cell, which will be described later on, the narrow band spectrum of the original pulse is absorbed, and only the broad spectrum during the ter- mination is transmitted.

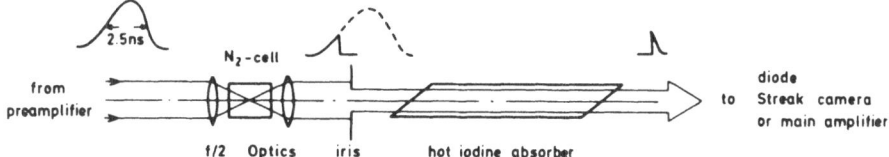

Fig.7 Setup of the FID-experiment using a 3 ns laser pulse

The resulting pulse duration is determined by
a) the duration of breakdown,
b) the dephasing time of the absorber and its absorption, whichever is long. In case of an ideal switch (termination infinitely fast) a pulse of half width T_2/α. L is obtained, where T_2 is the dephasing time of the absorber and α the absorption per cm and L its length. With typical T_2 times of 800 ps and values of α. L of 10 pulses with a duration around 80 ps are conceivable. The time history of the pulses obtained in Garching are shown in Fig.8. They were pro-

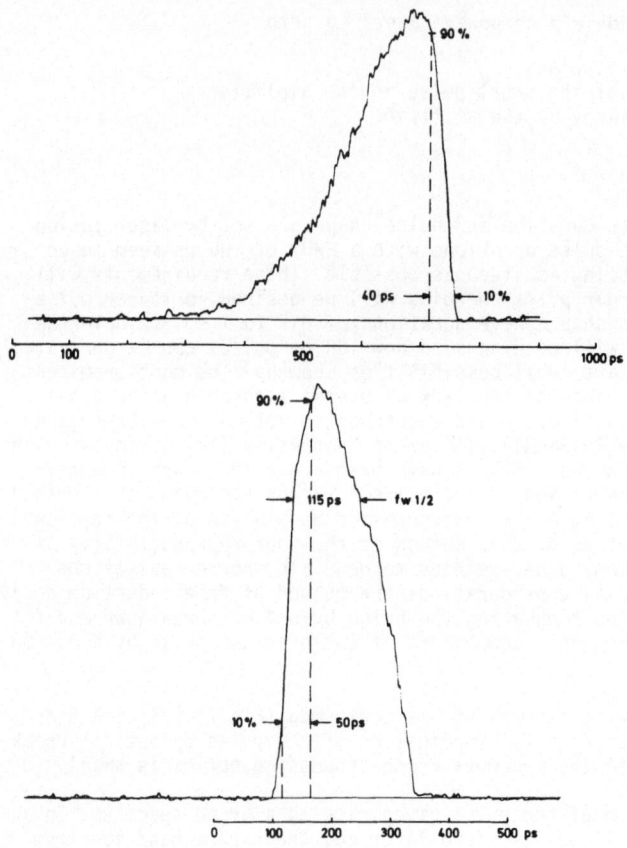

Fig.8 Time history of the FID-pulses
a) pulse form after the breakdown
b) pulse form after the narrow band pass filter

duced with an F/2 optics and have a duration of around 50 - 100 ps and an energy of $\sim$100 μ Joules. The duration is probably determined by the breakdown.
The disadvantage of this method are the unknown processes which occur in the
breakdown region.

A modification of the method consists in the use of a fast Pockels cell as the
switch instead of the breakdown. Pockels cells with rise times of 90 ps are
commercially available [16] . This scheme is depicted in the lower part of
Fig.9 where the breakdown is replaced by a fast gate. The sense of the Pockels
cell is not to cut out a part of the incoming pulse but to add a broad frequency spectrum to the incoming pulse by the fast risetime of the Pockels
cell. The narrow frequency spectrum is then eliminated by the narrow band pass
filter of the I_2-cell, which will be described in the next section. To get this
scheme to work a powerful narrow band pulse has to be generated, which can
easily be obtained using a low pressure short length oscillator. In Fig.9 an
experimental setup is depicted with which we got a smooth narrow band 50 ns

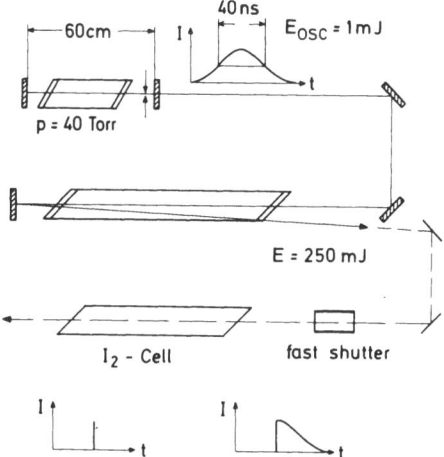

<u>Fig.9</u> Upper part: Setup for getting a smooth laser pulse (40 ns FWHM 250 mJ)
The lower part shows the proposed experiment using a fast Pockels
cell to get a FID pulse

laser pulse with a peakpower of 5 MW. The oscillator had a cavity length of
60 cm and a pressure of 40 Torr. The mode spacing was 250 MHz while the bandwidth of the laser medium was 600 MHz. This was enough to get only one longitudinal mode to lase. It should be mentioned that normally the oscillator
works in the gain switch mode, therefore the absence of any switching component
like a Pockels or a Kerr cell. The oscillator was amplified in a double pass
amplifier which gives an effective amplification of 300. When such a pulse is
cutted by a fast Pockels cell and is then passed in a fast narrow band absorber, an effect similar to the above described breakdown effect should arise,
resulting in a pulse duration around 100 ps, produced electronically and therefore probably more reproducible and with unchanged good divergence.

5. Pulse Propagation in an Amplifier Chain

Before discussing the problems connected with the storage of energy in the
amplifier first the propagation of a pulse in an amplifier chain will be considered. For this discussion let us assume we have already a series of amplifiers which can amplify the incoming pulse to the desired value. One of the
most critical points in laser plasma experiments is the energy contained in
so-called prepulses. This energy heats the target before the main pulse hits
it. Several effects have to be considered, as displayed in Fig.10 for a mode-
locked oscillator.
a) Energy leaking through the pulse cutting system
b) Background radiation passing the open pulse cutting system (precursor)
c) Amplified spontaneous emission of the amplifiers (fluorescence).

   The leaking energy can - hopefully - be sufficiently reduced by applying
several Pockels cells in series. However, to reduce the precursor to sufficiently low values only a saturable absorber can be used. Amplified fluorescence plays no essential role in case of the iodine laser because the Einstein
A cofficient is small and pumping times are short.

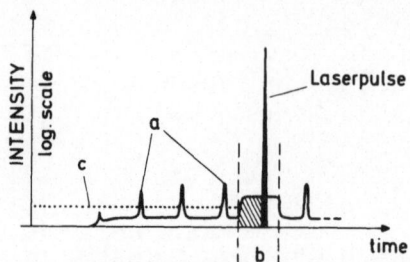

Fig.10 Effects, which lead to prepulses in a high power laser driven by a modelocked oscillator

A saturable absorber for the iodine laser can be designed by dissociation of molecular iodine in a heated cell [17, 18] . This has first been shown by V.A. Gaidash and others [19] . The additional requirement for a saturable absorber, namely, that the relaxation time of the upper level be long compared to the pulse width is fulfilled in case of the iodine cell. Conditions relating to the I_2 cell have been comprised in Fig.11.

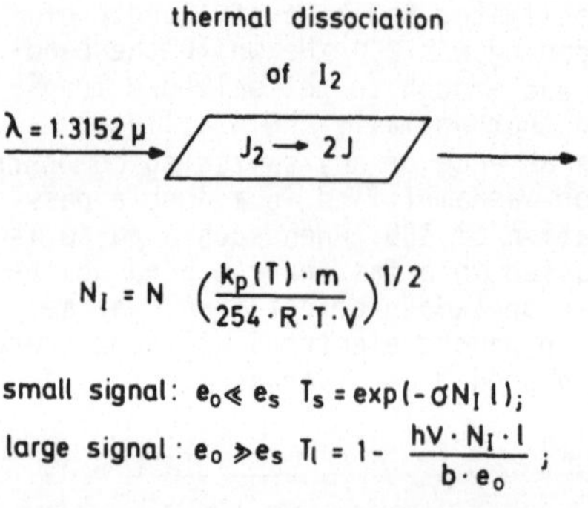

thermal dissociation

of I_2

$$\lambda = 1.3152\,\mu \longrightarrow \boxed{J_2 \rightarrow 2J} \longrightarrow$$

$$N_I = N \left(\frac{k_p(T)\cdot m}{254\cdot R\cdot T\cdot V}\right)^{1/2}$$

small signal: $e_0 \ll e_s$ $T_s = \exp(-\sigma N_I\, l)$;

large signal : $e_0 \gg e_s$ $T_l = 1 - \dfrac{h\nu\cdot N_I\cdot l}{b\cdot e_0}$;

Fig.11 Thermal dissociation of I_2 in a hot cell

The number density of iodine atoms is controlled by the temperature of the iodine cell. The cross-section for absorption depends mainly on the line width of the transition. As in the case of the laser it is adjustable by adding a foreign gas or by changing the temperature.

For the iodine cell to act as a saturable absorber its parameters have to be set to fulfil the condition

$$\left.\frac{\sigma\cdot b}{F}\right|_V \ll \left.\frac{\sigma\cdot b}{F}\right|_A$$

where σ_V is the cross-section for stimulated emission of the laser and σ_A

is the cross-section for absorption of the absorber. F is the cross-section of the beam in amplifier and absorber respectively, b is the degeneracy factor.

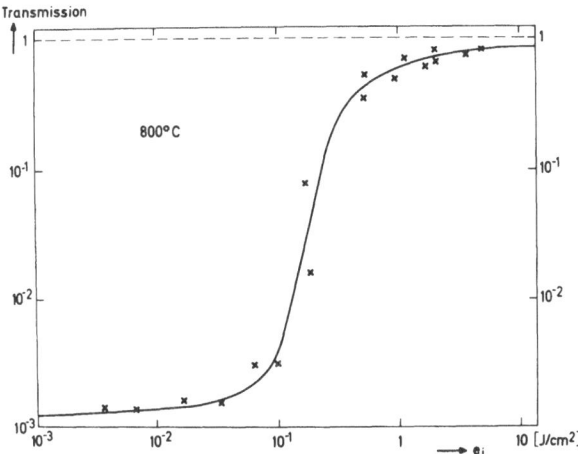

Fig.12 Behaviour of the hot I_2-cell for ns pulses
T_S = Small signal transmission
T_L = Large signal transmission

For ns pulses we found a behaviour displayed in Fig.12. As can be seen the iodine cell is well saturable with energy densities of a few Joules/cm². For the contrast ratio (small signal transmission / large signal transmission) values of 10^{-3} and better are obtained. Suppression of prepulses improved by a factor of 1000 could be demonstrated experimentally. Only the future can tell if these results can also be realized for 100 ps pulses.

So far, discussion has centered upon conventional methods of generation of nanosecond pulses. A good deal of the problems would be overcome, if the FID method could be taken into operation. The narrow frequency spectrum of a 100 ps pulse could easily be filtered out with an I_2 cell with no need of saturation for the main pulse. Contrast ratios of 10^{10} seem to be entirely feasible according to calculations.

The importance of good prepulse suppression becomes particularly evident if one considers pulse propagation in an amplifier system, which shows considerable saturation. The effects arising from saturated amplification are from special interest because all high efficient laser systems have to operate in the saturated mode. Otherwise nearly all the stored energy would remain in the amplifier. Dr. Olsen [20] investigated theoretically the influence of prepulse suppression, hyperfine relaxation [21] and saturation on pulse propagation in an iodine laser system called Plasterix which is shown in Fig.13. It differs from the one described in the literature [22] by the addition of Faraday rotators and a saturable iodine absorber. The theoretical model is based on the semiclassical laser equations because coherent effects have to be taken into account. The numerical results for two different cases are given in Fig.14.

134

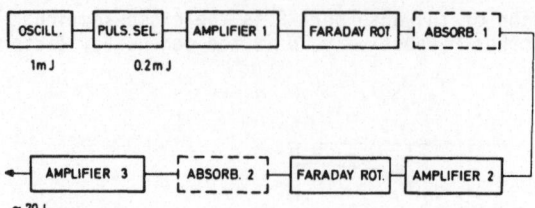

Fig.13   Schematic setup of a 80 J iodine laser

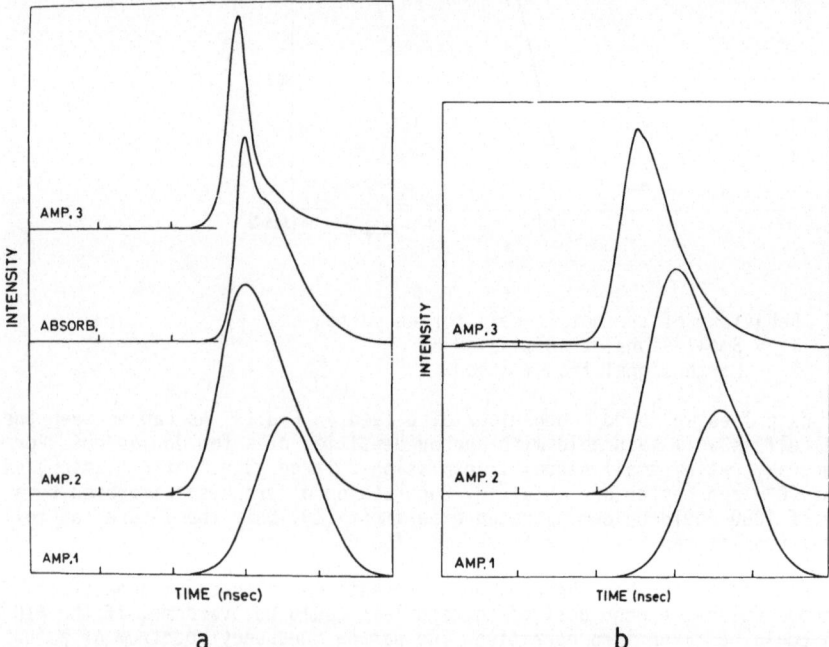

a   b

Fig.14   Propagation of the pulse through the amplifier chain (calculated
         pulse forms)
         a) with absorber (I_2-cell)
         b) without absorber

These two cases differ by the assumptions made for the prepulse. If the incoming pulse of 1 ns has a prepulse of 3 ns duration with a contrast ratio of
$1:10^4$, one obtains the pulse shape shown in Fig.14b. The Gausian pulse shape
of 1 ns half width is changed very little. At the exit we find a pulse duration of .76 ns and 84 Joules of energy. Because of saturation effects, the
pulse is steepened a little towards the front. Particularly noticeable, however, is the growth of the prepulse whose intensity is now 2 % of that of the
main pulse. The pulse shapes, however, change drastically if one includes a
saturable absorber in the amplifier chain. In Fig.14a one sees a considerable

steepening of the pulse. The halfwidth here is about 340 ps and the risetime about 200 ps. The contrast ratio is about 3×10^4. If one includes two such absorbers, the contrast ratio would improve to 1:10^7, a value quite satisfactory for laser target interaction experiments. The absorber reduced the total pulse energy by about 10 %, which is more than compensated for by the power increase. Preliminary experiments with this system show that saturation effects play a very important role. More precise measurements have to be conducted to compare the experiments with the calculations quoted.

An often asked question in the iodine laser business relates to self focussing problems. Several contributions may lead to a degradation of the beam quality [23] . They can be distinguished in two classes. The first leads to a static or intensity independent distortion of the wavefront, the second to a dynamic or intensity dependent one. Static distortions can in principle be corrected by a lens (although practically it might not be very easy to do). The main source of this kind is due to the following mechanism. Because of spatially inhomogeneous absorption of the pump light by the parent molecules a temperature gradient and thereby a transverse pressure gradient is built up in the mixture which sets the gas into motion what then changes the gas density and simultaneously the index of refraction. In the ASTERIX III system this effect causes phase variations of 1 up to 1.5 waves across the aperture summed up over the entire amplifier chain. Reduction can be achieved either by more homogeneous pumping or by using a buffer gas with a much higher molecular weight than that of Argon, SF_6 f.i.

The dynamic or intensity dependent distortion of the wavefront is caused by two sources. The first is a time integrating change of the refractive index that accompanies the transfer of excited iodine atoms to the ground state by stimulated mission. It can only occur when the laser medium is operated in saturation. It has the consequence that portions of the pulse (in time and space) see different indices of refraction what increases the focal spot diameter. This intensity dependent change of the refractive index can be expressed as the sum of two terms one counting for the different polarizabilities of the iodine atoms in the upper and the lower state (nonresonant contribution) and the other for the anomalous dispersion (resonant contribution). According to our estimates both termes can cause wavefront distortions which are clearly below one wavelength and are therefore of little importance, especially when they counteract each other.

The second contribution to be dealt with can result from the nonlinear index of refraction n_2. For the various gases encountered in the iodine laser (CF_3I, C_3F_7I, Argon, SF_6, Air) the n_2 value is at least 10^3 smaller than that of the Bk-7 glass. Even at the large distances (~ 100 m) which the pulse has to travel in gas laser systems the contribution of the gas to the beam break up integral is negligible. The B-value can therefore be calculated by considering only the glass components in the system. In ASTERIX III for instance (500 J, 500 psec throughout the system) one arrives at a value of 1.3 based on the average intensity which is close to the peak intensity because of the almost rectangular intensity profile in the regions of interest. This figure shows that n_2 does not play the limiting role in iodine as it does in glass laser systems. At this time we can say the divergence of the beam can be controlled within 5 times the natural divergence of the beam even for the highest powers of 1 Terawatt available today.

6. Scalability and Efficiency

For the construction of a high power laser with several or even 100 kJ pulse energy its scalability is the most important characteristic. This means that the laser can technically and physically scaled up to big diameter and high energy content. Large diameter of the lasers are necessary because the power flux density through the laser windows are limited due to nonlinear absorptions. For example a 100 kJ laser would need a diameter of more than 1 m if the critical power flux density would be 10^{10} W/cm^2 in 1 ns a value which has not been realized yet.

The number of the necessary amplifiers in an amplifier chain is determined by the self oscillator threshold for a single amplifier. This value is for instance 10^2 - 10^3 for the iodine laser. Normally one has to amplify a mJ-pulse up to kilojoules which means a total amplification of at least 10^6 is necessary. At least, because one wants to extract energy from the amplifier and therefore has to work in the large signal regime. As an example the laser ASTERIX III [24] has a total amplification of 10^9 and needs therefore 4 amplifiers.

Nearly the whole energy of a high power laser pulse stems from the last amplifier. According to this the total chain can be divided into the preamplifier and the main last amplifier the efficiency of which almost exclusively determines the total efficiency. The further discussion will be restricted to this main amplifier which should be excited by UV-light. Even if there are other excitation mechanisms like electron collisions in discharges [25] or the double photon excitation as proposed by Carman and Rhodes [26] the UV-Photolysis by flashlamps is the most developed and most frequently used method. Other UV-light sources as breakdown, wire explosions [27], arcs and so on, are not so flexible and reproducible up to now.

With regard to the flashlamps, there is a special difficulty: the shock wave problem. In ordinary iodine laser devices the gas is usually enclosed by a quartz tube which itself is surrounded by a couple of flashlamps (see Fig.15). If the gas is flashed a shockwave is generated which is probably caused by the burst of an iodide gas layer adsorbed at the wall of the quartztube. This shockwave runs with 1 - 2 times the velocity of sound into the laser gas which is a mixture of Argon and C_3F_7I. In this case typical shockwave velocities are 3 - 5 mm/10µs . The shockwave disturbs the homogeneity of the gas, so that the divergence of the beam gets worse and worse if the beam interacts with the shockwave. This means the region of the gas behind the shockwave is useless and lowers the efficiency of the system. As a consequence the gas has

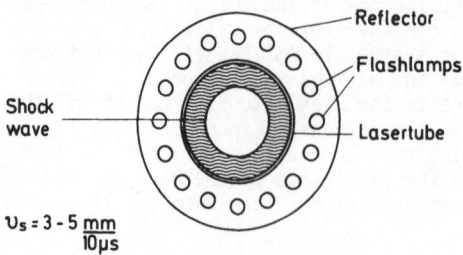

$v_s = 3 - 5 \dfrac{mm}{10µs}$

Fig.15　Schematic cross-section of an iodine amplifier

Fig.16 ASTERIX III end amplifier

to be flashed pretty fast on a time scale of 10 - 20 μs. Moreover flash lamps must be used which are expensive and which have to be run with a high energy load.

In the ASTERIX III end amplifier, which is shown in Fig.16, 64 lamps are necessary, through which 250 kJ in 20 μs are discharged. These lamps surround the quartz tube which has an inner diameter of 17 cm. The tube is roughened on the inner side to prevent an imaging of the lamps by the tube. The maximum efficiency of this device was measured by oscillator measurements. At a partial

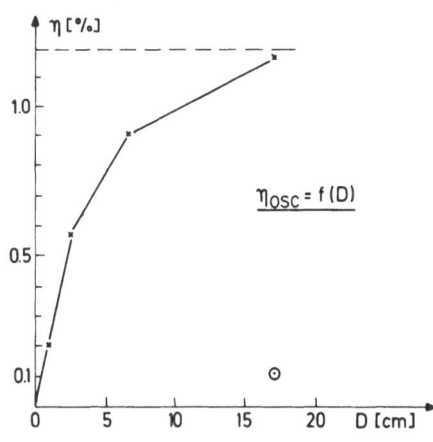

Fig.17. Efficiency of the iodine laser as a function of tube diameter (oscillator measurements)

pressure of 16 Torrs iodide it was 1,2 %. Similar measurements were performed for different tube diameters and the results are shown in Fig.17 in which the efficiency as a function of the diameter is plotted. In these measurements the disturbance by the shockwave plays no role, so we find the maximum efficiency by extrapolating the curve to large diameters to be about 1,3 %.

Now, how does this efficiency relate to the efficiency of the same device driven as an amplifier. In the diagram the amplifier efficiency is also given for the 17 cm amplifier, this means the efficiency drops by nearly a factor of 10 to 0,15 %. How can this decrease be explained? It is not the transition itself because it is homogenously broadened and the energy can be extracted at a single frequency in very short times. The reasons are summarized in Fig. 18. At first, we can not operate the amplifier with the same pressure as the oscillator because of the homogeneity of the inversion which should be equal within ± 15 %. The reduction of the pressure to 6 Torrs yields an efficiency reduction of 70 %. The imaging of the lamps by the tube causes hot spots. To avoid these high inversion zones the tube has to be roughened as already mentioned. The accompanied scattering losses reduced the efficiency by 20 %. The shockwave described above runs with a velocity of 5 mm/10 µs towards the axis of the tube thereby requiring that a ring of 1 cm thickness is shielded from any interaction with the laser beam passing through the tube 15 µs after the ignition of the flashlight. For longer time periods between ignition and passage of the pulse an even broader ring has to be shielded. The reduction by this shielding is about 20 %, taking into account that the highest inversion is located near the walls. An additional reduction is also related to the shockwave. The pulse has to pass through the amplifier when the lamps still pump, otherwise the ring has to be too broad. This reduction amounts 20 %, and we have reached an overall efficiency of about 0.4 %. The last reason for a further reduction stems from the laser transition itself, and is of fundamental

reason		α_i
homogenous inversion	lower pressure	$\alpha_1 = 0,7$
hot spots	etching	$\alpha_2 = 0,8$
shock wave	shielding	$\alpha_3 = 0,8$
shock wave	early pass	$\alpha_4 = 0,7$
bottle necking		$\alpha_5 = 0,66$
energy extraction		$\alpha_6 = 0,6$

$$\eta_{Amp} = \alpha_1 \cdot \alpha_2 \ldots \alpha_6 \cdot \eta_{osc} ;$$

$$\alpha_1 \cdot {}_2 \ldots \alpha_6 = 0,12$$

Fig.18 The various effects yielding a decreas of the efficiency of the iodine laser

nature. The laser transitions ends at the ground state which cannot be empted out during the laserpulse of 1 ns. This bottlenecking effect together with the fact that the upper level is twofold degenerated and the lower level fourfold gives a maximum extraction efficiency of 66 %. But this can only be achieved when the amplifier is totally driven in the saturation regime. In the case of ASTERIX III laser the input energy was 20 Joules. This gives together with the small signal amplification of thousand an energy extraction of 60 %. All these factors together decrease the total efficiency to 0.15 %.

This discussion has shown that the reduction of nearly one order of magnitude is a combination of several small effects each of them lying between 20 and 30 %.

Now, what can be done to eliminate these detrimental effects. Only one of them is from principal nature: the bottlenecking effect. But only this effect would give us a decrease of the total efficiency to 0.6 - 0.8 % depending on the degree of saturation. For the solution of these problems several proposals have been made [28] . The XeBr-excimer-flashlamps have been proposed as an excitation source for the iodine laser because of their probably high efficiency, defined as UV-radiation in the pump band of the iodide molecule over electrical input energy [29] . The flashtime of XeBr-excimers is in addition so short that the influence of the shockwave can be minimized. A disadvantage of this type of lamps is probably the short penetration depth of their wavelength because it coincides with the peak absorption of iodides, and therefore would cause inhomogeneities of the inversion. This can be overcome by bleaching effects but than the lamps have to be very powerful. In addition the chemistry of the laser gas for these high energy densities is not known. But nevertheless bleaching pumping with high efficient excimer lamps seems to be a very attractive solution. We want to propose another scheme which is not as radical and where we can use our normal lamps or even lamps with longer flashtimes. This solution would apply a neutral gas blanket instead of the quartz tube. Neutral gas means a gas which will not be activated by the lamps and which will not react with the walls. There are probably several possibilities to reach this goal. One possible realisation is shown in Fig.19. The neutral gas is flowing through the tube in the laminar streaming mode. Before the gas Argon enters the laser chamber the laser gas is added in the center of the stream. Before the iodide has reached the tube by diffusion processes it will be flashed. A density profile of the iodide can be established by applying several ring nozzles.

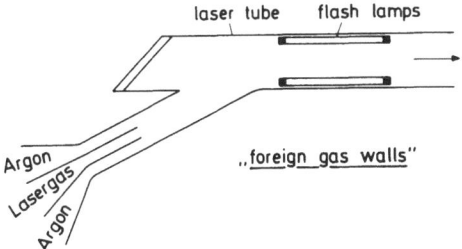

Fig.19 Device for establishing foreign gas walls to prevent the contact of the laser gas with the tube

This concept puts the difficulties from the quartz tube to the installation of a suitable flow system. But it increases the efficiency by a factor of 3 to 4 and has the additional advantage that it can be scaled to large diameters. In addition in this concept the long lifetime of the inversion could be applied and the flashlamps have not to be driven so hard. In principle the capacitor of Nd-glas systems could be used.

7. Summary

The discussion has shown that the potential of the iodine laser for producing high power laser pulses has not yet been exhausted. There are a variety of possibilities for a qualitative improvement.

The main advantages of the iodine laser compared with Nd-glass laser are the higher repetition rate and higher self focusing power level. If these advantages can be combined with new pumping devices an improvement of the efficiency to more than 0.5 % could be achieved, even with conventional flashlamps being used today.

References

1. J.V.V. Kaspar, G.C. Pimentel: Appl. Phys. Lett. 5, 231 (1964)
2. V.S. Zuev, V.A. Katulin, V.Yu. Nosach, O.Yu. Nosach: Sov. Phys. Jetp 35, 870 (1972)
3. W. Fuß, K. Hohla: Z. Naturforsch. 31a, 569 (1976)
4. F.T. Aldridge: IEEE QE-11, 215 (1975)
5. V.S. Zuev, V.A. Katulin, V.Yu. Nosach, O.Yu. Nosach: ZETF 62, 1673 (1972)
6. W. Fuß, K. Hohla: Report IPP IV/67 (Max-Planck-Institut für Plasmaphysik, Garching, Germany 1974)
7. F.T. Aldridge: Appl. Phys. Lett. 22, 180 (1973)
8. V.Y. Zaleskii, T.I. Krupenikova: Opt. Spectrosc. 439 (1973)
9. R.J. Donovan, D. Husain: Ann. Rept. Chem. Soc. London Ser. A 68 (1971)
10. H.J. Baker, T.A. King: J. Phys. D: Appl. Phys. 9, 2433 (1976)
11. E.P. Jones, M.A. Palmer, F.R. Franklin: Optical and Quantum Electronics 8, 231 (1976)
12. H.J. Baker, T.A. King: J. Phys. E 9, 287 (1976)
13. K. Hohla, W. Fuß, R. Volk, K.-J. Witte: Opt. Comm. 13, 114 (1975)
14. E. Yablonovitch: Phys. Rev. A 10, 1888 (1974)
15. E. Fill, K. Hohla, G.T. Schappert, R. Volk: Appl. Phys. Lett. 29/12, 805 (1976)
16. W.E. Martin, B.C. Johnson, K.R. Guinn, W.H. Lowdermilk: Laser Focus June 1977, 44
17. E.Fill, K. Hohla: Opt. Comm. 18, 431 (1976)
18. H.J. Baker, T.A. King: J. Phys. D 10, 169 (1976)
19. V.A. Gaidash, G.A. Kirillov, S.B. Kormer, S.G. Lapin, V.I. Shemiakin, V.K. Shirigin: ZETF Lett. 20, 243 (1974)
20. J.N. Olsen: Journ. Appl. Phys. 47/12, 5360 (1976)
21. E.A. Yukov: Sov. J. Quant. Electr. 3, 117 (1973)
22. K. Hohla, G. Brederlow, W. Fuß, K.L. Kompa, J. Raeder, R. Volk, S. Witkowski, K.-J. Witte: J. Appl. Phys. 46, 808 (1975)
23. K.-J. Witte: Reports PLF 2 and 3 (Projektgruppe für Laserforschung der Max-Planck-Gesellschaft, Garching, Germany 1977)
24. G. Brederlow, K.-J. Witte, E. Fill, K. Hohla, R. Volk: IEEE QE-12, No. 2, 152 (1976)
25. L.P. Pleasance, L.A. Weaver: Appl. Phys. Lett. 27, 407 (1975)

26. R.L. Carman, Ch.K. Rhodes in: Advanced Laser Systems for laser driven thermonuclear fusion applications, submitted to Springer Verlag K.G.
27. N.G. Basov, V.S. Zuev: Il Nuovo Cimento 31 B, 129 (1976)
28. K. Hohla, K.-J. Witte: Report IPP IV/90 (Max-Planck-Institut für Plasma-physik, Garching, Germany 1976)
29. W.F. Krupke, E.V. George: UCRL-Report No. 77523 (Lawrence Livermore Laboratory, USA 1976)

ASTERIX III, A Terawatt Iodine Laser

K.J. Witte, G. Brederlow, K. Eidmann, R. Volk, E. Fill, K. Hohla,
and R. Brodmann

Projektgruppe für Laserforschung
der Max-Planck-Gesellschaft zur Förderung der Wissenschaften e.V.
8046 Garching, FRG

1. Introduction

In this paper the progress will be outlined which we made with our high power
laser system ASTERIX III in the last months. This system will be used for fusion experiments. We shall start with a brief description of the system, then
continue with the latest results and conclude our paper with some remarks on
a number of effects causing a wave front distortion and thereby leading to a
degradation of the beam quality.

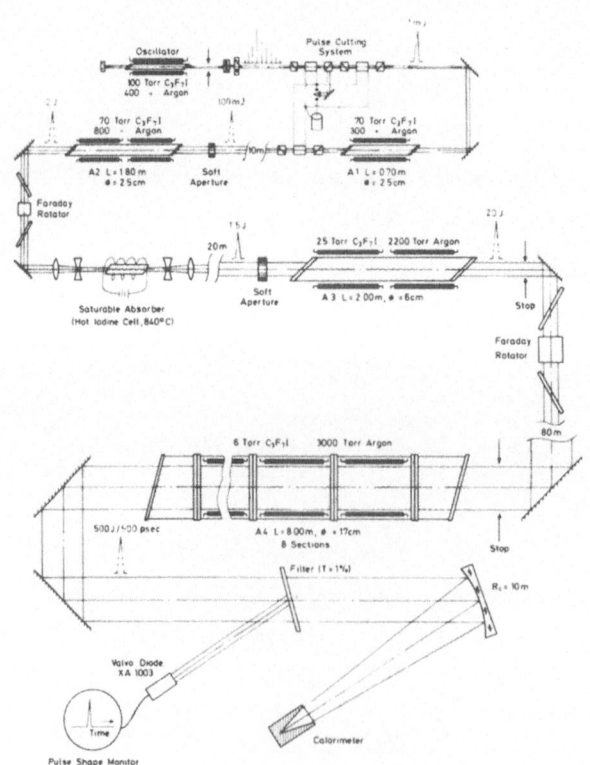

Fig.1 Schematic of ASTERIX III

2. Description of the System

The schematic of ASTERIX III is shown in Fig.1. Special care has been taken of the pulse cutting system whose contrast ratio is as high as the small signal amplification of the entire amplifier chain which is usually between 10^9 - 10^{10}. The oscillator and all the amplifiers are connected to a gas recycling system which removes the quenching constituents (I_2) from the gas and replaces the fraction of the C_3F_7I used up in the photolysis process. Thus a repetition rate of 1 shot each 8 minutes is possible. The essential energy gain takes place in the last amplifier which is 10 m long and equipped with 64 flashlamps (for more details see [1]).

Although there is a long distance between the individual amplifiers it turned out that this isolation by distance is not sufficient to allow for a sustainable small signal amplification of 10^{10}. For this purpose we incorporated two Faraday rotators and a saturable absorber thereby also protecting the laser system against the light backreflected from the target.

3. Results

3.1 Power Output and Prepulse Suppression without a Target

The first shots we made with the system just described were done without a target. About that time the quartz tube of the last amplifier had not yet been etched. Under these conditions we got the intended power of 1 TW (500 J, 500 ps FWHM) in several shots at an overall efficiency of 0.15 %. The pulse shape did not stay constant throughout the amplification process. The most substantial pulse steepening and shortening occurred in the saturable absorber and to a less degree in the last two amplifiers.

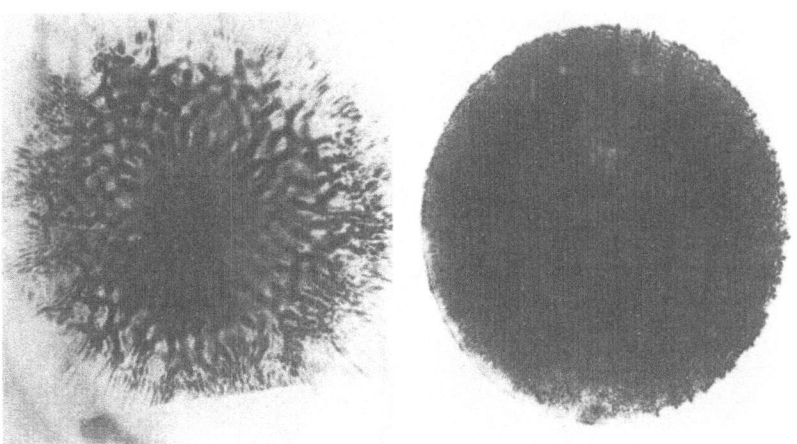

Fig.2 Burn pattern at the exit of the 4th amplifier
       Left: Unetched tube; shock wave not shielded
       Right: Etched tube; shock wave shielded

144

In these shots the prepulse energy was about 50 mJ. The main contribution results from the prelasing of the amplifiers (spontaneous emission $\sim$ 100 µJ, leak rate of the pulse cutting system $\sim$ 10 µJ) which is externally caused by amplifier coupling (spikes of $\sim$ 100 ns duration). Besides other tests this could also be proven by the fact that the incorporation of additional stops in the amplifier chain with diameters chosen such that the zones of extremely high inversion density were blocked reduced the prepulse energy to the sub-millijoule level and improved the beam quality considerably because the interaction of the beam with the shock wave was now also avoided (see Fig. 2).

3.2 Preliminary Target Experiments

After these encouraging results we started our first target experiments. They are preliminary in the sense that they were made at a moderate power level and the diagnostic was still incomplete. Its schematic is shown in Fig. 3. The first 15 shots we have made so far have demonstrated that with a target the total prepulse energy can be brought to the same submillijoule level as without a target. These results are really a big step forward because it was originally thought that the prepulse problem could be a severe limitation upon the performance of a high gain iodine laser working in the saturation regime.

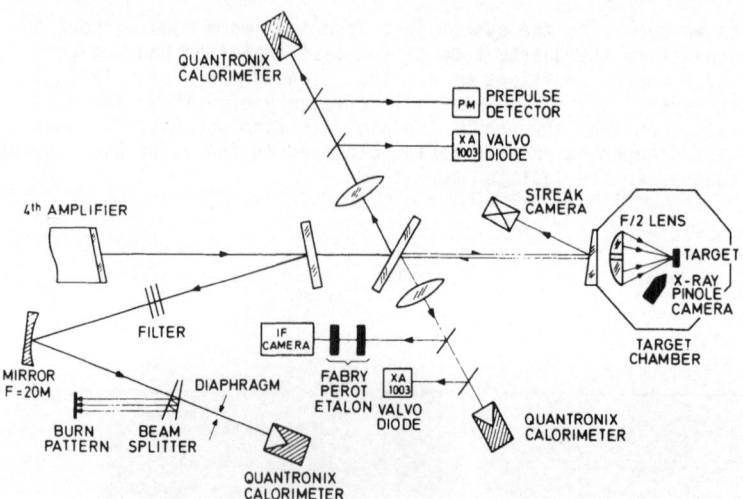

Fig.3 Schematic set up of the diagnostic used

The energy measurements of the light backreflected through the lens yielded similar results as those already obtained with the smaller PLASTERIX laser [2] . The reflection coefficient was between 10 - 20 % at a power density level of several times 10^{15} W/cm^2 on the target. In the PLASTERIX-experiments 40 % was observed, but there a F/1 optics was used. The backreflected light was broadened to a width of 15 Å compared with the width of 0.1 Å of the incoming light and probably also shifted, but to what extent could not be measured yet. The strong broadening of the backreflected light has the effect that it cannot destroy the laser because only a very small fraction of it falls inside the amplification bandwidth of the iodine transition.

Fig.4 shows on the left a picture of the plasma taken with the X-ray came-
ra (resolution 15 μm; radiation used ≳ 1.3 keV). The plasma diameter (the
little white spot in the center) is about 70 μm. According to the beam quality measurements (see Fig.5) the laser beam is 5 times diffraction limited
leading to a focal spot diameter of 30 μm by an ideal F/2 lens. Unfortunately,
the lens used in our set up is not free from errors; it has a diffraction
disk of 40 μm the ideal value being 6 μm. How the lens errors combine with
the wave front distortion is difficult to say. If they are additive a spot
diameter of 70 μm results which is a reasonable value compared with the
plasma diameter of also 70 μm.

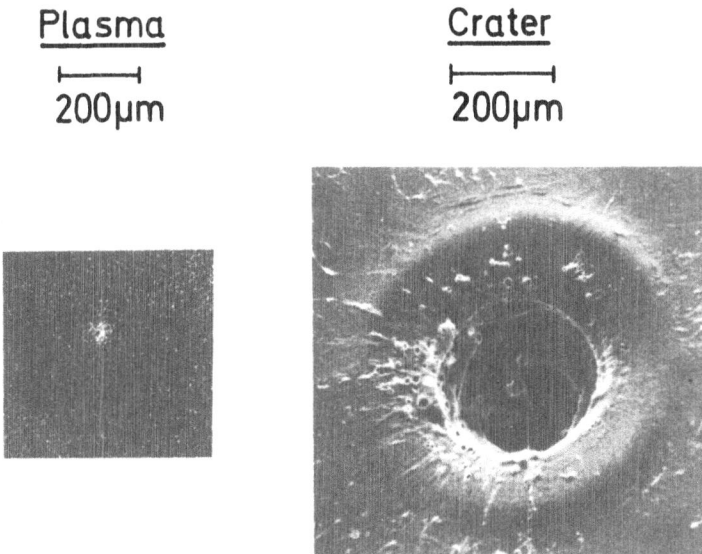

Plasma   ## Crater
⊢————⊣   ⊢————⊣
200μm   ### 200μm

Fig.4 Plasma diameter and crater shape of a plexiglass target

On the right of Fig.4 the crater is shown generated by the laser beam in
a plexiglass target (taken with a scanning electron microscope). For an assess-
ment of the beam quality the crater diameter which must be much larger than
the focal spot diameter is not as important as the shape of the crater, especially that of its rim. From the comparison of experiments made with Nd-lasers
delivering similar pulses it can be concluded that the iodine beam is indeed
clean and that its wave front is at least free from major distortions.

3.3 Beam Quality Measurements

The beam quality measurements were done by determining the focussibility with
the multiple spot- and diaphragm method (see Fig.3) at various positions in
the amplifier chain (without a spatial filter yet). The result is summarized
in Fig.5 where the ratio of the measured focus spot size and the ideal spot
size is plotted versus the beam path length. Whereas the beam quality degradation is rather smooth and modest up to behind the 2nd Faraday rotator it

is more distinct at the exit of the 4th amplifier. We ascribe this unexpected reduction of the beam quality mainly to the imperfect quality of the optical components like mirrors and windows lying between the 3rd and 4th amplifier. This conclusion could be drawn from the fact that the beam quality did not change regardless whether the 4th amplifier was fired or not.

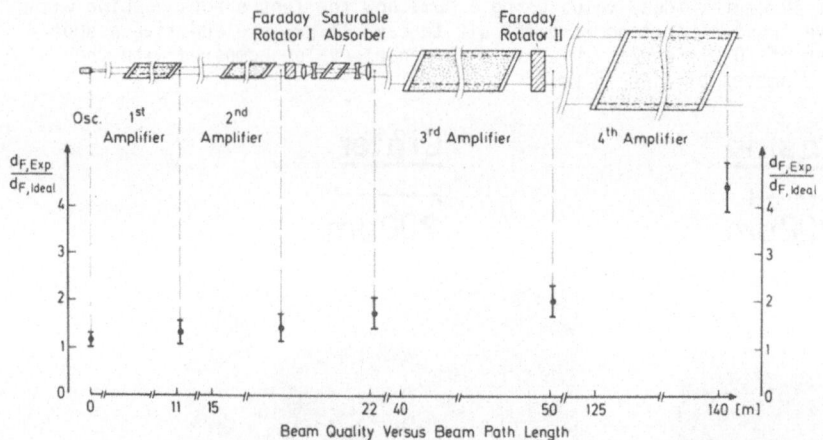

<u>Fig.5</u> Beam quality at various positions in the laser system

4. Sources of Wave Front Distortion

Besides these deficiences there are some other sources connected with the gaseous laser medium itself which may lead to a wave front distortion and thereby to a poor focussibility of the laser beam. Two groups are distinguished. The first leads to a static or intensity independent wave front distortion, the second to a dynamic or intensity dependent one. Static distortions can in principle be corrected by a lens although practically it might not be quite easy to do. The main source of this kind is due to spatially inhomogeneous pumping. It causes a temperature gradient and thereby a pressure gradient which sets the gas into motion what then changes the gas density and simultaneously the refractive index. We have strong experimental and theoretical evidence that this effect leads to a phase variation of 1 up to 1.5 waves across the aperture of the 4th amplifier in the ASTERIX III system. This distortion - although still tolerable - can be minimized by using a buffer gas with a high molecular weight (SF_6 instead of Argon) and by more homogeneous pumping [3] .

The dynamic wave front distortion is caused by two sources. The first is a time integrating change of the refractive index that accompanies the transfer of excited iodine atoms to the ground state by stimulated emission provided the laser medium is operated in saturation. It has the consequence that the leading edge of the pulse sees a different refractive index than the trailing edge thereby changing the focussing properties. It can be shown that this effect can be expressed as the sum of two terms one counting for the different polarizabilities of the iodine atoms in the upper and the lower state and the other for the anomalous dispersion. According to our estimates both terms can cause phase variation across the aperture at the exit of the 4th amplifier which are clearly below one wave so that they are of little importance, especially when they counteract each other.

The second contribution to the dynamic wave front distortion can result from the nonlinear refractive index n_2 of the glass components in the laser system. The n_2 of the laser medium itself and of the air ($\sim$120 m path length in air) is roughly 1000 times smaller so that it is of no importance. We have therefore calculated the B-integral in ASTERIX III considering only the glass components. At an exit power of 1 TW (500 psec throughout the system) we arrived at a value of B = 1.3. Even if an additional beam splitter and the focussing lens is included (10 cm more of glass) B would not increase above 2.6. For good focussibility a value of 5 should not be exceeded [4] . The remaining gap is still large enough to allow the incorporation of additional optical components if necessary.

From all the results and considerations presented above we may conclude that the iodine laser is very well suited for high power applications and moreover a system scalable to large sizes delivering short pulses in the 10 to 100 kJ range.

5. Literature

1. K. Hohla: (invited paper)"High Power Lasers and Applications"Conference, 20 - 22 June 1977, Munich

2. K. Eidmann, Chr. Dorn, R. Sigel: IPP-Report IV/95 (1976)

3. K. Witte, PLF-Report 2 (1977)

4. J.C. Guyot, A. Bettinger, D. Auric, Congress "Moyens Technique De La Fusion", Paris, 6 - 10 Dec. 1976

A Simple High Energy TEA CO Laser

W.E. Schmid

Projektgruppe für Laserforschung
der Max-Planck-Gesellschaft zur Förderung der Wissenschaften e.V.
8046 Garching, FRG

1. Introduction

The pulsed carbon monoxide laser is a very promising candidate for laser induced chemistry because it is emitting laser light on many lines between 4.8
and 5.8 μm. In this spectral range many interesting chemical compounds can be
vibrationally excited as e.g. ketones, olefines, imines and metal carbonyls.

In addition the CO laser has a theoretical efficiency of 90 %. The only
relevant loss is the V-T relaxation associated with the anharmonic V-V pumping. Efficiencies of 60 % have been demonstrated by Mann et al [1]in an electron beam controlled discharge laser (EDL) at cryogenic temperatures, and laser energies of 1.2 kJ have been achieved by Center [2] in a similar but big-
ger device.

In contrast to the CO_2 laser where high laser energies have been achieved
with TEA lasers as well as with EDL's no high energy TEA CO lasers have been
reported to date. Cohn reported 0.1 mJ for a resistor array laser [3] and
100 mJ from a photoinitiated discharge laser [4] . Champagne [5] achieved 35
mJ from a helical resistor array laser. All this lasers have been operated at
cryogenic temperature. At room temperature Jeffers and Wiswall [6] reported
0.84 mJ from a resistor array laser.

Laser action in electrically initiated chemical CO lasers (CS_2/O_2) has
been reported too. Ahlborn et al [7] achieved several watts and Ahl and Binns
[8] 115 mJ. The goal of our work was to modify our 40 J CO_2 laser which is
described in [9] in order to yield substantial energies of CO laser output.

2. Apparatus and Experimental Technique

The cross section of the laser is shown in Fig.1. The discharge takes place
between two Rogowski shaped electrodes. These electrodes are made of plastic
in a casting mold and then electroplated with 100 μm of copper and a thin layer
of nickel. The active volume is 5 x 5 x 50 cm^3 = 1.25 l. The sliding spark
array which is producing the UV light for the preionization is inserted in the
cathode. The UV light penetrates through a. wire mesh into the active volume.
The anode and the incoming laser gas are cooled by a cold ethanol or liquid
nitrogen flow. Electrodes and active volume are surrounded by polyurethane
foam for thermal insulation inserted in a lucite chamber.

The resonator consisting of a gold coated 10 m mirror and a coated germanium flat with R = 97 % or 99 % reflectivity is mounted directly on to the

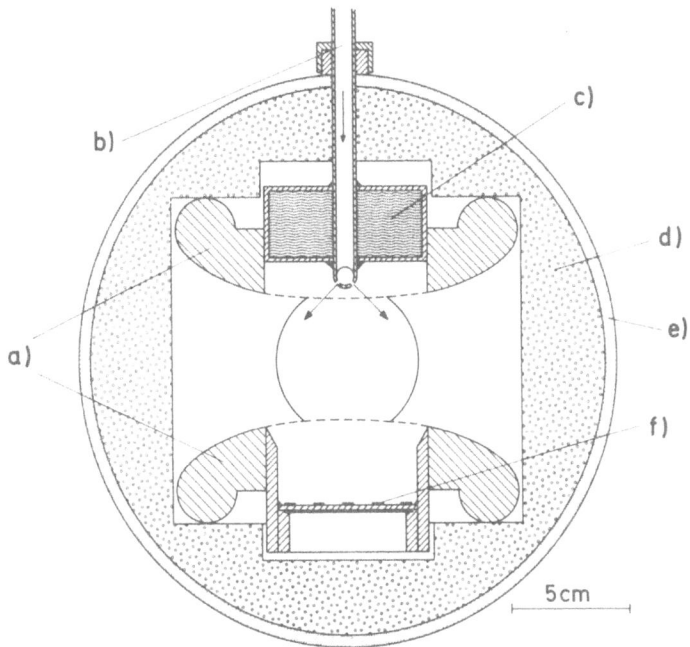

Fig.1 Cross section of the laser chamber. a) Rogowski electrodes, b) gas in-
let, c) cooling bath, d) thermal insulation, e) housing, d) UV-preionization unit

laser housing. The outcoupling mirror can be warmed up by a weak stream of
warm air to prevent the condensation of moisture. The laser gas consists of
a mixture of He, N_2, and CO and a small amount of tri-n-propylamin as a seed
gas for the UV-preionization. The commercial purity CO is flown through a
cooled molecular sieve (Zeolith 5 Å) to remove mainly metal carbonyls which
are detrimental to the discharge [2] .

Pulseshapes have been measured with a gold doped germanium detector, energies with a Gentec ED 200 or ED 500 joulemeter. For the extracavity selection
of single lines a McPherson 0.3 m monochromator has been used.

3. Results

Operation at cryogenic temperature in other CO lasers yielded the best results by
- enhancing the population of higher levels by the anharmonic V-V pumping
  process
- narrowing the rotational distribution.[1]

[1]This is important as all papers except [7] report that there is only partial
 inversion of single rotational states and not of vibrational states.

Attempts to cool the laser with liquid nitrogen, failed due to the fact, that all organic compounds which could serve as seed gases for the preionization have practically zero vapour pressures at this temperatures. Consequently no distributed glow discharges with high stored energies could be established. It was decided then to return to moderate temperatures (-20°C) maintained by an ethanol flow and an electric refrigerator. Here we got stable discharges with up to 280 J of stored electrical energy.

By optimizing the gas mixture for maximum energy output we could increase laser energy up to 5 J. Fig.2 shows the dependence of the laser energy W and the efficiency η on the stored electrical energy W_{EL}. Maximum efficiency is 2 % which is of course far from theoretical and even from values reached in EDL's. But it is about twice as high as the values of Cohn in a cryogenic photoinitiated TEA CO laser [4] . The energy could not be further increased due to discharge instabilities.

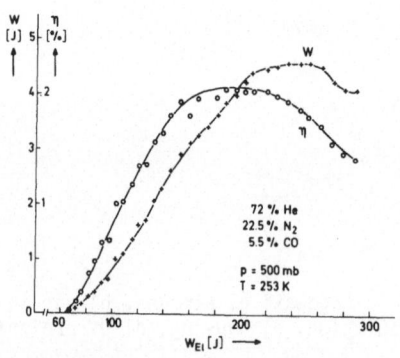

Fig.2 Dependence of laser energy W and efficiency η on stored electrical energy W_{EL}. Total gas pressure p = 500 mb, T = 253 K

Fig.3 Dependence of laser energy W on gas temperature t °C . Reflectivity of the outcoupling mirror R = 99 %.

The laser energy is considerably enhanced even by slightly cooling the laser gas as is shown in Fig.3. Further cooling was not possible without ruining the discharge.

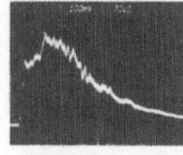

p = 530 mb
t_H = 72 µs

20 µs/div

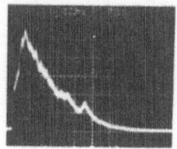

p = 960 mb
t_H = 32 µs

Fig.4 Total pulse shape at two different pressures

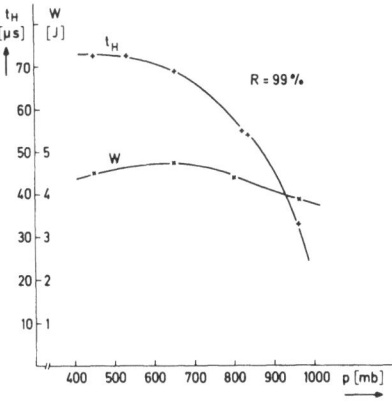

Fig.5 Dependence of the energy W and pulse width t_H (fwhm) on the gas pressure

The pulse shape of the total pulse is shown in Fig.4 for two different pressures. Pulse width decreases rapidly with increasing gas pressure due to faster V-V transfer by collisions. As energy shows only a weak dependence on gas pressure, laser power reaches 100 kW at high pressures.

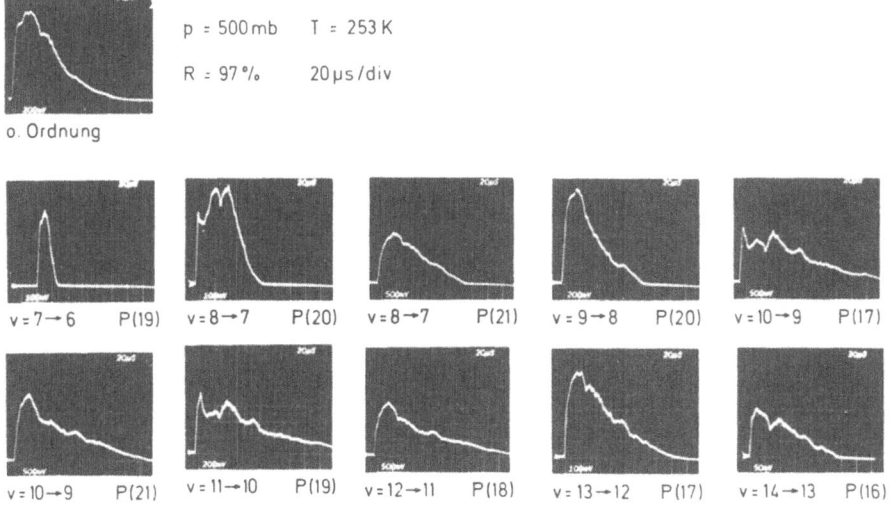

p = 500mb   T = 253 K

R = 97 %   20 μs/div

o. Ordnung

$v = 7 \rightarrow 6$ P(19)	$v = 8 \rightarrow 7$ P(20)	$v = 8 \rightarrow 7$ P(21)	$v = 9 \rightarrow 8$ P(20)	$v = 10 \rightarrow 9$ P(17)
$v = 10 \rightarrow 9$ P(21)	$v = 11 \rightarrow 10$ P(19)	$v = 12 \rightarrow 11$ P(18)	$v = 13 \rightarrow 12$ P(17)	$v = 14 \rightarrow 13$ P(16)

Fig.6 Pulse shape of the total pulse and of the single vibrational rotational transitions

Whereas Fig.4 shows the pulse shape of the total pulse Fig.6 shows in addition the shapes of the single lines which this pulse is composed of. Different delay times, pulse shapes and durations are the result of the complicated kinetics of the CO laser. Vibrational energy of the molecules is cascading down by stimulated transitions the higher states being populated by collisional processes. P-transitions from v = 7 → 6 to v = 15 → 14 with J varying from 23 to 12 have been observed and are plotted in Fig.7.

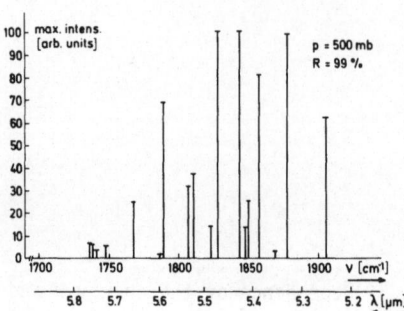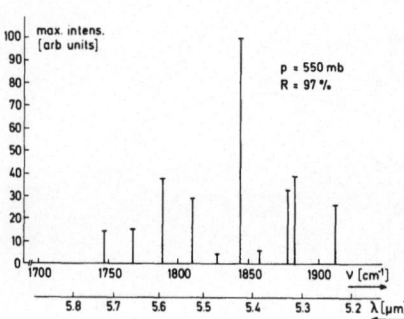

<u>Fig.7</u> rel intensity of the single vibrational rotational transitions at different resonator losses and slightly different pressures[2]

It is worth mentioning that increasing pressure and resonator losses reduce the number of lines considerably without affecting energy too much in the measured range, so that one can hope to concentrate the laser energy on a few lines. Intracavity line selection has to our knowledge only been tried once for pulsed CO lasers [6] . At the pressures used there - 30 - 40 torr - it was not possible to increase energy on single lines with a intracavity grating. We hope that in the pressure range employed here this will be possible.

4. Conclusion

By optimizing the parameters we have increased the energy of TEA CO lasers from the 100 mJ range to 5 J in a small simple device working at a convenient temperature. By increasing gas pressure and choosing suitable resonator parameters it has been demonstrated that the laser energy can be concentrated on a few lines.

5. Acknowledgement

The author is gratefully indebted to K.L. Kompa for initiating this work and W. Fuß for many helpful discussions. The excellent technical assistance of F. Aigner is also acknowledged with thanks.

6. References

1. M.M. Mann, D.K. Rice, and R.G. Eguchi: IEEE J. Quant. Elect. QE-<u>10</u>, 682 (1974)
2. Robert E. Center: IEEE J. Quant. Elect. QE-<u>10</u>, 208 (1974)
3. D.B. Cohn: Appl. Phys. Lett. <u>8</u>, 343 (1972)
4. D.B. Cohn: IEEE J. Quant. Elect. QE-10, 459 (1974)
5. L. Champagne: Appl. Phys. Lett. <u>23</u>, 158 (1973)

[2]the high intensity at 1842 cm^{-1} may be due to a accidental coincidence of P(23) v = 9 → 8, P(17) v = 10 → 9, and R(11) v = 14 → 13

6. William Q. Jeffers and Charles E. Wiswall: IEEE J. Quant. Elect. QE-$\underline{7}$, 407 (1971)
7. B. Ahlborn, P. Gensel, and K.L. Kompa: J. of Appl. Phys. $\underline{5}$, 2487 (1972)
8. Jeffrey L. Ahl and W.R. Binns: IEEE J. Quant. Elect. QE-$\underline{12}$, 26 (1976)
9. W.E. Schmid: IPP Report IV/84 (1975)

Band Width of an Oscillating CO_2 Laser Transition

W.J. Witteman and A.H.M. Olbertz

Department of Applied Physics, Twente University of Technology
Enschede, The Netherlands

By using mode-lock techniques for CO_2 laser pulses one can observe that the pulse width at atmospheric pressure desreases with increasing helium percentage. This is a striking phenomenon because a rigorous analysis of the pulse-forming mechanism yields an expression in which the pulse width is inversely proportional to the square root of the band width, whereas absorption measurements on laser-gas mixtures have shown that the effect of helium to the line broadening is considerably less than those of the molecular components [1, 2]. Thus one might expect an increasing pulse width with the helium concentration. This inconsistency is the subject of the present paper.

We shall describe an experimental technique for measuring the band width of a laser transition under the actual conditions and consequently we shall apply this method to a TEA CO_2 laser. The results show clearly that under laser-discharge conditions the band width increases with increasing helium percentage. Using these data excellent agreement between observed and calculated pulse widths as a function of helium percentage can be obtained.

In the case of active AM mode locking with the modulation frequency very close to the axial mode spacing it is possible to treat the mode locking in terms of individual modes and to obtain analytic expressions for the pulse width and the maximum detuning [3]. The maximum detuning can also be expressed in terms of the change x of the distance between the mirrors from the position of maximum performance. So we found for the pulse width:

$$\tau_p = \frac{1}{\pi} (2 \ln 2)^{\frac{1}{2}} \left(\frac{2G_0}{m\alpha_a}\right)^{\frac{1}{4}} \left(\frac{1}{\Delta\nu_N \, \nu_m}\right)^{\frac{1}{2}}, \tag{1}$$

where G_0 is the singel-pass power gain, m and α_a are modulation parameters of the mode locker, $\Delta\nu_N$ is the band width and ν_m is the modulation frequency. For the maximum positional shift of the mirrors we found:

$$x_m = \frac{c}{\pi \, \Delta\nu_N} \left[G_0^* \left\{G_0^* - L_0 - \alpha_a(1 - m)\right\}\right]^{\frac{1}{2}}, \tag{2}$$

where G_0^* is the small-signal gain of the medium and L_0 is the single-pass power loss. For small values of $\alpha_a(1 - m)$ the distance x_m decreases linearly with $\alpha_a(1 - m)$, so that the maximum positional shift for the lossless mode locker (m = 1) can easily be obtained by extrapolating the observations at small values of $\alpha_a(1 - m)$ to those for zero modulation. It is seen from (2) that x_m is independent of the cavity length and for weak modulation also independent of the modulation depth. For a practical CO_2 laser system operation at one atmosphere we estimate from (2) that x_m will be a few centimeters. This indeed has been observed.

We employ a TEA laser with two identical Rogowski-shaped electrodes at a distance of 1.7 centimeter, trigger wires and a small concentration of the low ionization seed gas, tri-n-propylamine [4]. The discharge length is 80 cm and the discharge is fed from a two-stage Marx generator with a total capacitance of 0.02 µF and a charge voltage of 60 kV. The cavity is formed by one totally reflecting mirror with a radius of curvature of 5 meter and a germanium flat with the outer surface AR coated. There is a small angle of $2 \cdot 10^{-3}$ radians between the two surfaces of the flat, so that interference effects between the surfaces are absent.

The mode locking occurs by means of active mode locking with an acousto-optic device to produce the required periodic loss modulation. The modulator consists of a germanium crystal which was cut in such a way that the optical beam impinged on the crystal at Brewster's angle. Is was assured by outlining the system that the refracted radiation inside the cavity meets the phase grating at the Bragg angle. The modulator was driven by the attached lithium-niobate transducer, which was connected with a tunable RF power oscillator. The acoustic resonance frequency spacing of the transducer was measured as 5.0 MHz and that of the germanium crystal as 34 kHz with a resonance width of 12 kHz. The long term frequency stability of the RF oscillator was within 10 KHz. The modulation frequency at 40.028 MHz of the modulator was therefore stabilized within 10 KHz. In order to suppress off-axial modes a diaphragm of 5 mm diameter was just behind the mode locker.

The observed pulse train consisted of about 20 to 50 pulses with a duration of about 250 nanoseconds. At maximum detuning the train contained only a few pulses. The pulses in the centre of the train were practically equal and any variation in pulse width near the centre could not be observed. By detuning the laser system for fix parameters of electrical input energy and modulation depth the pulse form was always well-defined and the width measured at half-intensity points was observed to change very little. Only the pulse height decreased remarkably with the frequency shift, but did not decrease continuously to zero. At a certain frequency shift the mode locking and oscillation disappeared abruptly. In Fig. 1 we have plotted the maximum positional shift x_m as a function of the modulation depth for various gas compositions at one atmosphere total pressure. From the figure we see that x_m depends only little on α_a. It is also observed that x_m decreases with increasing helium percentage. It was observed that the points with positive and negative x_m values were located symmetrically within a few percent with respect to $x = 0$.

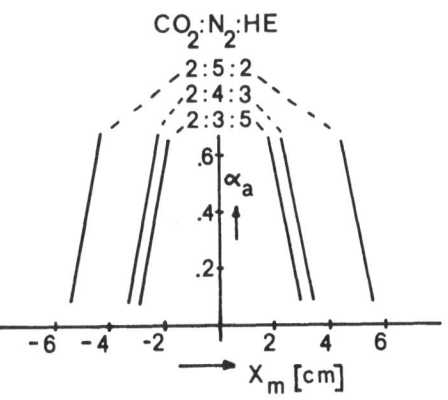

Fig. 1 Maximum positional shift of the mirror distance as a function of the modulation depth

From the observation we can easily extrapolate to values of x_m for $\alpha_a \rightarrow 0$, so that according to (2) the line width can be obtained independently from the modulation parameters. The small-signal gain G_0^* can be obtained from gain measurements with a low-power cw laser. We observed that for our discharge conditions the variation of G_0^* with the helium concentration in the range of interest was more or less constant. For a (2:3:5) CO_2, N_2, He mixture the small-signal gain was 2.0% per centimeter, so that $G_0^* \cong 1.6$, about the same as for the (2:3:2) and (2:3:10) mixtures. Using this value for G_0^* and calculating $L_0 = 0.60$ from the outcoupling loss, diapragm and absorption through the modulator, we deduced by means of (2) the line width as a function of the helium percentage. The results are plotted in Fig. 2.

For comparison we have also plotted (dotted curve) the band width as expected from absorption measurements under neutral conditions. Further we measured the pulse width as a function of the helium percentage by means of a hot-hole detector having a time constant of 200 picoseconds. The results are plotted in Fig. 3. It is clearly seen that the pulse width indeed decreases with the helium percentage, consistently with the observation that the line width increases with the helium percentage. The measurements in Fig. 3 are compared with the calculations according to (1) using the measured line widths indicated in Fig. 2.

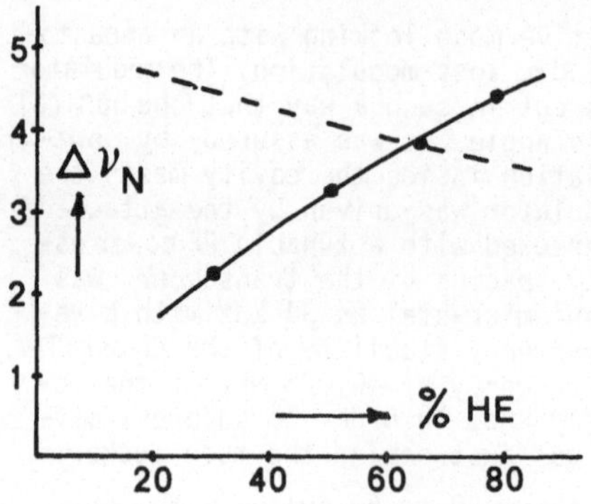

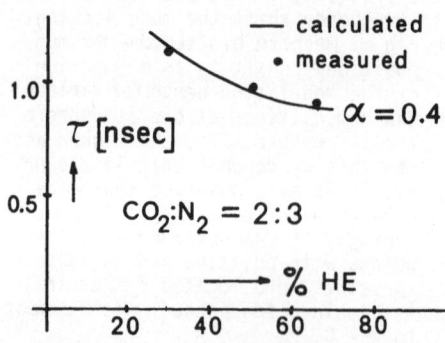

Fig. 2 Band widths calculated from the maximum frequency detuning and from absorption measurements (dotted curve)

Fig. 3 Pulse width versus He concentration for an atmospheric gas mixture

References

1. R.R. Patty, E.R. Manring and J.A. Gardner; Appl. Optics 7, 2241 (1968)

2. O.R. Wood; Proc. IEEE 62, 335 (1974)

3. W.J. Witteman and A.H.M. Olbertz; IEEE J. Quantum Electr. QE-13, (1977)

4. B.J. Reits and A.H.M. Olbertz; Appl. Phys. Letters 26, 335 (1975)

Penning Ionization in Doped CO_2 TEA Lasers

B.J. Reits

Department of Applied Physics, Twente University of Technology
Enschede, The Netherlands

1. Introduction

Low-ionization impurities have a profound influence upon CO_2 laser gas discharge parameters. It is our purpose to describe a discharge mechanism that
accounts quantitatively for this influence. The discharge mechanism is based
upon the assumption that metastable gas molecules in the discharge are able
to ionize the seed gas. This is a so-called Penning reaction; it is schematically given by

$$M^* + S \rightarrow M + S^+ + e, \tag{1}$$

where M^* is a colliding particle in a metastable state ans S a seed-gas molecule. Apart form Penning ionization the model also deals with direct ionization, attachment and recombination. We will arrive at two differential
equations that succesfully predict the discharge behaviour

2. The Penning model

Examination of possible processes in the laser-gas components, i.c. helium,
nitrogen and carbondioxide, shows that only metastable nitrogen can be involved. BORST [1] has shown that for low-energy electrons only $A_3\Sigma_n^+$, the
$a_1\pi_g$ and the $E^3\Sigma_g^+$ states are significantly populated. The $A_3\Sigma_n^+$ state needs
considerable additional vibrational energy in order to be able to ionize our
seed gas, tri-n-propylamine. The $E^3\Sigma_g^+$ state is of little importance compared to the $a_1\pi_g$ state, due to its small cross section and its high threshold. In our calculation we therefore only consider the $a_1\pi_g$ state. For this
state the cross section is available. We used this cross section for calculating the excitation coefficient α' of the $a_1\pi_g$ state in our reference
gas, a 1:7:30 mixture of CO_2, N_2 and He. This excitation coefficient α' is
shown in Fig. 1 together with the ionization coefficient α for our gas mixture.

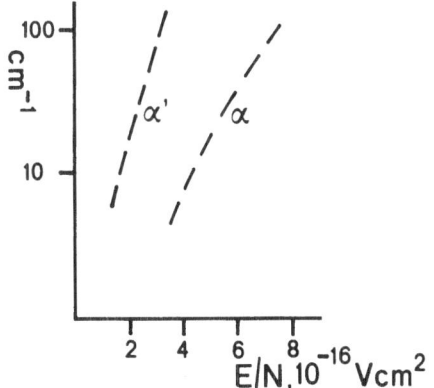

Fig. 1 Ionization coefficient and
excitation coefficient α' in a
1:7:30 mixture of CO_2:N_2:He

It can be seen that the production of metastables is about two orders of magnitude higher than the production of electron-ion pairs by direct ionization. This means that with an effective seed gas direct ionization might become unimportant as compared to Penning ionization.

The $a_1\pi_g$ state has a radiative lifetime of 115 μsec [2]. In the gas discharge its lifetime will be much shorter due to deactivating collisions. For our gas mixture the $a_1\pi_g$ lifetime τ_{gas} is about 2.7 μsec [3]. The addition of a low-ionization seed gas further reduces the lifetime. When we call the lifetime of the metastable due to only the seed gas τ_{dope}, we find for the real lifetime τ:

$$\frac{1}{\tau} = \frac{1}{\tau_{gas}} + \frac{1}{\tau_{dope}} . \tag{2}$$

We are able now to write down the equations that rule the doped discharge. The metastable concentration n_m is given by:

$$\frac{dn_m}{dt} = \alpha' \, n_e \, v_0 - \frac{1}{\tau} \, n_m, \tag{3}$$

where α' is the excitation coefficient of the N_2 $a_1\pi_g$ state, n_e the electron density and v_0 the electron drift velocity. The electron density n_e is given by:

$$\frac{dn_e}{dt} = (\alpha - a) \, n_e \, v_0 - \beta \, n_e^2 + \frac{k \, n_m}{\tau_{dope}} , \tag{4}$$

where α is the ionization coefficient, a the attachment coefficient, β the recombination coefficient and k a seed-gas efficiency term. In our modelling we use $\beta = 3 \cdot 10^{-7}$ cm^3 sec^{-1} and $k = 0.1$ [3]. It is possible now to solve (3) and (4) with the aid of the transport coefficients given in [4].

3. Application to laser constructions

Our basic laser construction consists of two Rogowski-shaped electrodes with a 50×1×1 cm^3 gas discharge, together with a 30 Ω series resistor. Photopreionization is obtained by two knife edges beside the anode. With this construction we can illustrate the stationary solution

$$(\alpha - a + \frac{\tau}{\tau_{dope}} \alpha') \, v_0 - \beta \, n_e = 0. \tag{5}$$

We therefore apply a 25 kV step voltage to the basic laser construction

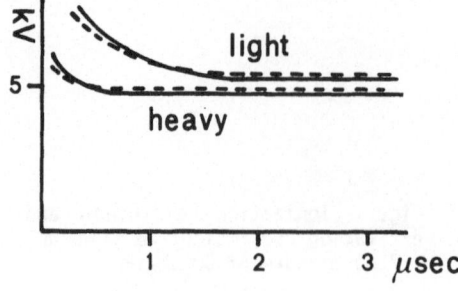

Fig. 2 Measured and calculated laser voltages when a 25 kV step voltage is applied to the basic laser construction, for a light and a heavy doping

and measure the voltage across the electrodes. Measured voltages are shown in Fig. 2 by the solid lines; the dotted lines give solutions of (3) and (4). We see that the laser voltage indeed stabilizes at a value that depends on the seed-gas concentration in the expected way.

The transient solution deals with the case where we start with a large concentration of metastables, produced by a peaking capacitor wired across the electrodes. The plasma density is given by:

$$\frac{dn_e}{dt} = \frac{n_{om} k}{\tau_{dope}} \exp(-\frac{t}{\tau}) - \beta \, n_e^2. \tag{6}$$

Solutions of this equation have been studied elsewhere [5]. Measured and calculated laser voltages for this case are shown in Fig. 3. The laser is

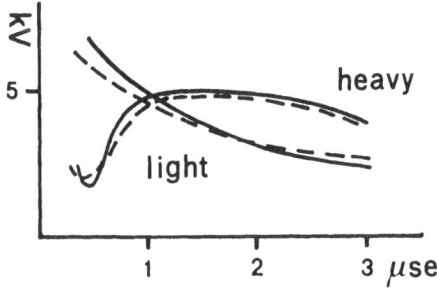

Fig. 3 Measured (solid) and calculated (dotted) laser voltages for a heavy and a light doping when a peaking capacitor is used

fed from a 0.1 µF capacitor and a 7200 pF capacitor is wired across the electrodes.

Finally we give the solution for the case where the laser is fed from a 0.1 µF capacitor but no peaking capacitor is used. Laser voltages are shown in Fig. 4. This laser configuration was used in our amplifier modules, now

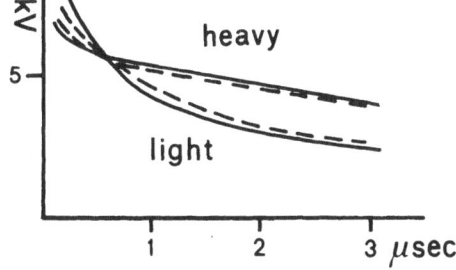

Fig. 4 Measured (solid) and calculated (dotted) laser voltages for a heavy and a light doping when no peaking capacitor is used

having apertures up to 12 cm. They have a 4%/cm small-signal gain and give output powers of 50 J/l in an oscillator configuration. They are further characterized by their perferctly homogeneous discharge.

References

[1]. W.L. Borst; Phys. Rev. A5, 648, 1972

[2]. W.L. Borst, E.C. Zipf; Phys. Rev. A3, 979, 1971

[3]. B.J. Reits; J. Appl. Phys., Sept. Issue 1977

[4]. J.J. Lowke, A.V. Phelps, B.W. Irwin; J. Appl. Phys. 44, 4664, 1973

[5]. A.H.M. Olbertz, B.J. Reits; Appl. Phys. Lett. 28, 200, 1976

Part IV

Frequency Conversion

Nonlinear Processes in the Infrared and Ultraviolet*

C.K. Rhodes

Molecular Physics Center, SRI International
Menlo Park, CA 94025, USA

1. Introduction

The field of nonlinear optics, which began with the seminal work of FRANKEN, HILL, PETERS, and WEINREICH [1] in 1961, has now grown to a mature technology. The earlier work centered on the use of ruby (694.3 nm) and Nd (1.06 μm) lasers, since these sources readily provided the optical intensities necessary for the observation of nonlinear phenomena in a wide variety of materials. These effects included second harmonic generation [2], self-focusing [3], and coherent effects [4]. In addition, new spectroscopic advances with dye lasers have extended nonlinear techniques to high resolution spectroscopic studies [5].

More recently nonlinear processes in the infrared have been observed for a wide range of phenomena including isotopically selective molecular dissociation [6], frequency conversion [7,8] and molecular spectroscopic analyses [9]. In this article we will explore the details of the infrared mechanisms as a prelude to the analysis of analogous processes in the ultraviolet. region.

2. Infrared Processes

Optical down-conversion of 10 μm radiation serves as an excellent example of the properties and flexibility of two-quantum processes. Furthermore, with the current availability of high power infrared sources, particularly at 10 μm, we expect a rapid extension of the practical range of nonlinear amplitudes for both pure scientific and applied purposes.

The two-quantum 16 μm laser demonstrated [8,10] in ammonia has several interesting properties generally common to systems excited by nonlinear mechanisms. Indeed, from an analysis of this case, several issues related to the utility of the optical degrees of freedom for these systems emerge. In this context we note, that in order to achieve efficient energy conversion, four conditions must be satisfied: (1) the establishment of the resonance conditions, (2) the creation of the proper absorption length, (3) the saturation energy requirement, and (4) a collisional constraint arising from col-

---
*Work supported by the U.S. Energy Research and Development Administration under Contract No. AT(04-3)-115.

lisional redistribution of the excited molecular energy. To properly satisfy these four conditions, we have at least six variables including the optical intensities of the two waves I_1 and I_2, the relative polarization of the two fields denoted as $\sin \varphi = \hat{\epsilon}_1 \cdot \hat{\epsilon}_2$, where $\hat{\epsilon}_1$ and $\hat{\epsilon}_2$ are the respective unit polarization vectors, the optical pulse length τ, the medium density ρ, and the length of the excited medium ℓ.

A. Optical Stark Shift

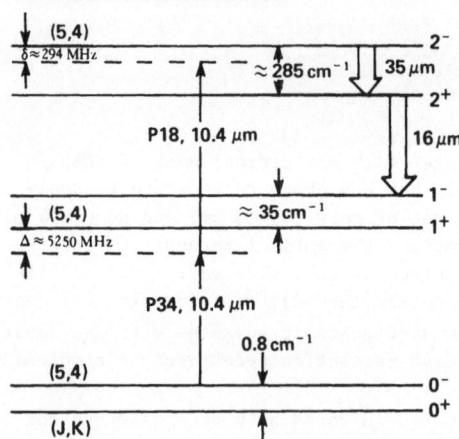

Fig.1 Partial energy level diagram of the ν_2 mode of $^{14}NH_3$ illustrating the two-photon absorption utilized to populate the $2\nu_2^-(5,4)$ state and the 35.50- and 15.88-μm transitions serving as examples of the generated laser wavelengths.

In connection with the first factor, resonance, we observe from Fig. 1 that in the absence of a DC Stark field, the two-photon resonance is 294 MHz off exact resonance. If strong coupling is to be achieved, this detuning must be significantly reduced by some means. This could be accomplished by broadening either the transition or laser linewidths, tuning the laser lines, or appropriately tuning the molecular resonance. Since broadening processes decrease the coupling strength and tuning of ∼ 300 MHz by conventional lasers is not feasible, we consider a novel approach for tuning the molecular energy levels into resonance in the absence of a DC Stark field by using the optical Stark effect [11,12].

The work of BISCHEL et al. [11] provides a demonstration of the optical Stark effect in NH_3 and shows that the $Q(5,4)$ two-photon transition illustrated in Fig. 1 can be tuned into resonance with a power density of approximately 1 MW/cm^2. This value is consistent with unfocused CO_2 TEA lasers. Therefore, it is possible to achieve the required resonance condition by illumination with the appropriate optical intensities, though it is necessary to integrate this condition with the other three constraints stated above, governing the coupling to the excited medium. For instance, efficient excitation in the length ℓ implies, for unfocused beams, that the intensity will change appreciably along the length of the medium. Although modest focusing could compensate for this aspect, we will consider in Section 2.D an alternate method that illustrates additional features of controlled optical coupling to the medium.

In the lowest order analysis, the expression for the optical Stark shift ΔE_n of the nth level is given by

$$\Delta E_n = \tfrac{1}{4} \sum_m \left\{ \frac{\left|\vec{\mu}_{mn} \cdot \vec{\xi}\right|^2}{E_n - E_m - \hbar\omega} + \frac{\left|\vec{\mu}_{mn} \cdot \vec{\xi}\right|^2}{E_n - E_m + \hbar\omega} \right\} \tag{1}$$

in which the subscript m designates the intermediate states, $\vec{\xi}$ is the optical electric field, ω is the optical frequency, and $\vec{\mu}_{mn}$ is the electric dipole transition moment. On account of a near resonance in the present case, the summation is dominated by a single term. We are then left with two terms,

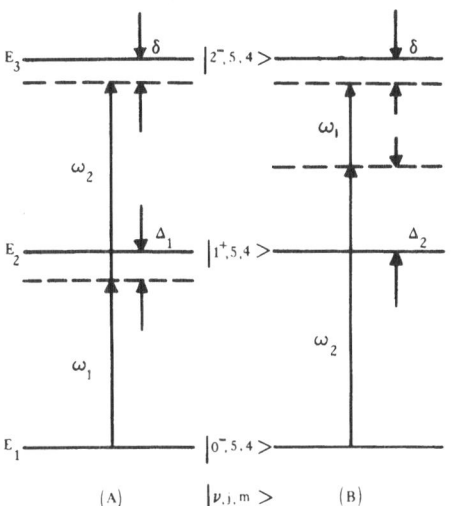

Fig.2 Graphical illustration of the two terms contributing to the amplitude for absorption.

corresponding to the pair of diagrams illustrated in Fig. 2. For the transition illustrated in Figs. 2A and 2B, the parameters [11] are $\delta = 294$ MHz, $\Delta_1 = 0.165$ cm^{-1}, $\Delta_2 = 14.9$ cm^{-1}, $\omega_1 = 1033.488$ cm^{-1}, and $\omega_2 = 1048.661$ cm^{-1}. With the definitions of parameters given in Fig. 3, the equations

$$E_3 - E_2 = -(\omega_1 + \Delta_1) \tag{2}$$

and

$$E_3 - E_2 = \omega_2 - \Delta_1 + \delta \tag{3}$$

hold. With $\delta/\Delta_1 \sim 10^{-1}$ and $\Delta_1/\Delta_2 \sim 10^{-2}$, if we ignore terms of the order 10^{-2}, we arrive at the simple estimates

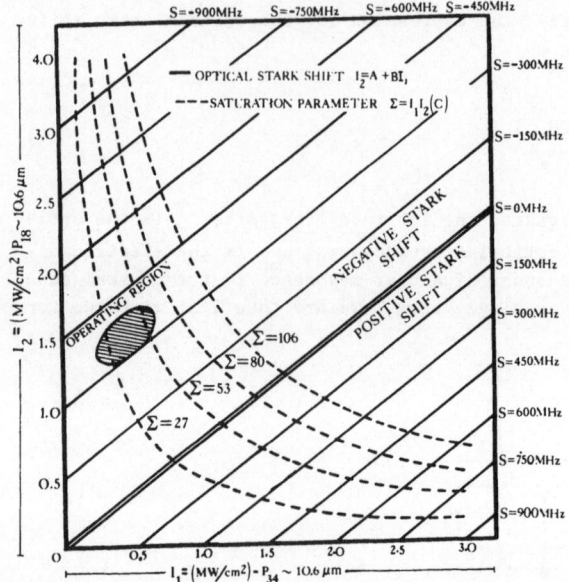

Fig.3 Illustration of the contours of constant Stark shift (S) and saturation parameter (Σ) in the I_1-I_2 plane. The operating region is indicated as the shaded ellipse.

$$\Delta E_1 \cong -\tfrac{1}{4} \frac{\left(\vec{\mu}_{12} \cdot \vec{\xi}_1\right)^2}{\Delta_1} \tag{4}$$

and

$$\Delta E_3 = -\tfrac{1}{4} \frac{\left(\vec{\mu}_{23} \cdot \vec{\xi}_2\right)^2}{\Delta_1 - \delta} \tag{5}$$

in which $\vec{\mu}_{ij}$ are the transition moments connecting their respective levels and $\vec{\xi}_1$ and $\vec{\xi}_2$ are the amplitudes of the two waves.

If we now define the total shift S as

$$S = \Delta E_3 - \Delta E_1 \tag{6}$$

we can now write

$$S \cong \frac{2\pi}{c} \left[\frac{\mu_{12}^2}{\Delta_1} I_1 - \frac{\mu_{23}^2}{\Delta_1 - \delta} I_2 \right] \tag{7}$$

in which c is the speed of light and I_1 and I_2 are the intensities of the waves at frequencies ω_1 and ω_2 respectively. All the parameters in (7) are known from previous spectroscopic studies [11]. This result states that I_2 is linearly related to I_1 for a constant shift S. Rewriting (7) we obtain

$$I_2 = \left(\frac{Sc}{2\pi}\right)\left(\frac{\Delta_1 - \delta}{\mu_{23}^2}\right) + \left(\frac{\mu_{12}}{\mu_{23}}\right)^2\left(\frac{\Delta_1 - \delta}{\Delta_1}\right) I_1, \tag{8}$$

an expression which we will utilize further in the discussion below.

B. Saturation

We now recall that another important condition is saturation of the two-photon transition. This condition is estimated by defining a linecenter saturation parameter (Σ) analogous to the single photon case [13] by

$$\Sigma = 2W_{fg} \, \tau > 1 \tag{9}$$

in which W_{fg} is the two-photon transition rate and τ is the upper state lifetime. Using standard order theory, W_{fg} can be written as [11]

$$W_{fg} = 8.44 \times 10^4 \, \frac{I_1 I_2 (W^2/cm^4)^2}{\Delta\nu_D} \, (P_{fg})^2 \tag{10}$$

where the effective matrix element [11] is $P_{fg} = \mu_{12}\mu_{23}/\Delta$ and $\Delta\nu_D$ is the two-photon Doppler width in MHz. Here μ_{ij} are the transition dipole matrix elements in debye and Δ is the detuning from the intermediate state in GHz. For $\tau \sim 3$ nsec (appropriate pressure broadened width), $\Delta\nu_D = 168$ MHz, $(P_{fg})^2 = 2.8 \times 10^{-5}$ and $I_1 I_2 = 0.75 \times 10^{12}$ (W^2/cm^4), we estimate that $\Sigma \cong 40$. This indicates that the two-photon transition would be completely saturated, if excitation were provided exactly at line center and that the energy absorption rate scales as the product of the two intensities $(I_1 I_2)$.

C. Combined Optical Stark Shift and Saturation Conditions

We are now in a position to combine the information contained in (8) and (9) by representing them in the I_1-I_2 plane as shown in Fig. 3. The straight lines represent loci of constant optical Stark shifts, while the hyperbolae denote contours of constant energy deposition Σ for line center irradiation. Since we wish to combine saturation with the proper Stark shift, the correct operating area is defined by the appropriate intersection of these contours as indicated in Fig. 3. These conditions have now been firmly established by experiment [10].

D. Polarization

Optical polarization can also play an important role. Since it is desirable
to preserve the correct Stark shift throughout the medium, we wish to remain
on the contour of constant $\Sigma = \delta$ parametrically as a function of distance z
along the medium. Recalling that saturation of the medium was a specified
condition for coupling, we know that the decrements in intensity ΔI_1 and ΔI_2
are linearly related to the distance z and that energy balance requires

$$\frac{\Delta I_2}{\Delta I_1} = \frac{\omega_2}{\omega_1} \; . \tag{11}$$

Therefore, if the slope of (8) matches the value given by (11) above, we
have the condition

$$\left(\frac{\mu_{12}}{\mu_{23}}\right)^2 \left(\frac{\Delta_1 - \delta}{\Delta_1}\right) = \frac{\omega_2}{\omega_1} \tag{12}$$

which will ensure a constant Stark shift in the saturated regime under conditions of appreciable absorption.

Two factors, frequency and polarization, can be used to establish the
equality demanded by (12). Because we regard the frequencies as very nearly fixed, we now examine the effect of polarization in establishing the required condition. The variation of the direction cosine matrix elements
provides a considerable range of adjustment of the parameters in (11). We
emphasize that the influence of polarization has been experimentally demonstrated in previous studies [9] on CH_3F confirming the presence of these
effects. Substitution of the appropriate values for the Q(5,4) transition
in $^{14}NH_3$ leads to the conclusion that the rotational direction cosine matrix
elements Φ_{ij} must satisfy the relationship

$$\frac{\Phi_{12}}{\Phi_{23}} = 1.482 \tag{13}$$

This condition can be established exactly by the correct choice of optical
polarization [10].

The discussion given above has illustrated the manner in which the available electromagnetic degrees of freedom can manifest themselves in the dynamics of many ·quantum amplitudes. Although the examples given involved vibrational systems in the infrared, precisely analogous mechanisms apply at
shorter wavelengths, namely, the ultraviolet. We now continue with an analysis of the observation of two-quantum processes in the ultraviolet region.

3. Ultraviolet Processes

A. Ultraviolet Sources

The recent development of intense discharge excited ultraviolet laser sources utilizing the rare gas halogen systems [351.1 nm (XeF*), 248.4 nm (KrF*), and 193.0 nm (ArF*)] are enabling the observation of new classes of non-linear phenomena in the ultraviolet. These sources are rapidly evolving as the laboratory analogues of the CO_2 laser in the ultraviolet region with current capabilities of $\sim$ 100 mJ/pulse and focused intensities of $\sim 10^9$ W/cm^2. One may predict confidently that this new technology will open many fresh avenues of fruitful activity in the study of electronically excited atomic and molecular species.

Multi-photon processes offer especially interesting possibilities, since a substantial quantity of energy can be deposited per target molecule (7 to 13 eV for a two-photon process) enabling the selective excitation of very highly excited states. Moreover, as in the infrared, these processes are quite general and can be applied to a wide class of atomic and molecular species.

B. Estimated Two-Photon Absorption Cross Sections

Estimates for two-photon processes can be readily made for several materials. Following the development furnished by BISCHEL et al. [9], which includes Doppler-free aspects and references to earlier material, we may write the cross section $\sigma(\nu_2)$ for the absorption of the wave at frequency ν_2 as

$$\sigma(\nu_2) = \frac{(2\pi)^3}{\hbar c^2} I_1 \nu_2 \left| M_{fg} \right|^2 g(\nu_1,\nu_2). \tag{14}$$

The derivation of (14) considered two waves with frequencies ν_1 and ν_2 and intensities I_1 and I_2, respectively. The factors appearing in (14) are the lineshape factor $g(\nu_1,\nu_2)$, which contains the linewidth of the transition, and the two-quantum matrix element M_{fg} which is written in the form

$$M_{fg}(\omega_1,\omega_2) = \sum_k \frac{< f |\hat{\epsilon}_1 \cdot \vec{\mu}_{op} |k > < k |\hat{\epsilon}_2 \cdot \vec{\mu}_{op} |g >}{E_{kg} - \hbar\omega_2} +$$

$$\frac{< f |\hat{\epsilon}_2 \cdot \vec{\mu}_{op} |k > < k |\hat{\epsilon}_1 \cdot \vec{\mu}_{op} |g >}{E_{kg} - \hbar\omega_1}. \tag{15}$$

The unit vectors $\hat{\epsilon}_1$ and $\hat{\epsilon}_2$ appearing in (15) denote the polarizations of the optical waves, $\vec{\mu}_{op}$ represents the electric dipole operator, and g, k, and f denote the ground, intermediate and final states, respectively. In the limit $\omega_1 \rightarrow \omega_2$, the case under present consideration, both terms in (15) become equal so that

$$M_{fg}(\omega_1,\omega_2) = 2\Sigma_k \frac{< f |\hat{\epsilon}_1 \cdot \vec{\mu}_{op} |k > < k |\hat{\epsilon}_1 \cdot \vec{\mu}_{op} |g >}{E_{kg} - \hbar\omega_1} . \tag{16}$$

This resulting matrix element contains the customary product of two transition moments over the appropriate energy denominator.

Estimates of two-quantum rates can be easily obtained by application of (14) through (16) provided that we have the relevant wavelengths, transition moments, and linewidths.

Table I illustrates the results of estimates for various atomic and molecular systems. Note that the range in susceptibility spans several orders of magnitude and that the estimates contained in Table I involve several diverse types of transitions including bound-bound processes in both atoms (Kr) and molecules (H_2) and bound-free transitions (Xe ionization, N_2O dissociation).

C. Preliminary Experimental Results

The estimated coupling parameters given in Table I suggest that two-quantum processes of this nature should be observable for several systems. Among the more likely cases would be the excitation of Xe* (5p6p) with 248.4 nm (KrF*) radiation. In particular, we describe below some preliminary measurements [14] involving two-photon absorption at 248.4 nm in the process

$$\gamma + \gamma + Xe \, (^1S_o) \rightarrow Xe \, (5p^5 6p[\tfrac{1}{2}]_o) \tag{17}$$

followed by the reaction

$$Xe* + N_2O \, (\tilde{X}^1\Sigma^+) \rightarrow Xe + N_2 \, (X^1\Sigma_g^+) + O(^1S). \tag{18}$$

The latter reaction leads to the emission of green light on the collisionally stimulated $O(^1S) \rightarrow O(^1D)$ auroral transition. Therefore, the green emission observed from Xe/N_2O mixtures serves at the signature of the absorption process given by (17).

A schematic of the experimental apparatus is shown in Fig. 4. A transverse-electric-discharge-pumped krypton-fluoride excimer laser was used of a type that has been described by Burnham and Djeu [15]. This source emits 20-30 mJ pulses of ~ 10 nsec duration.

The laser beam was focused into the experimental cell by a plano-convex Suprasil lens having a focal length of 50 mm. The area of the focal spot was ~ 0.2 x 0.3 mm, giving a peak focused power of ~ 3×10^9W. An f-1 Suprasil lens collected and collimated the fluorescence from the sample. This light then passed through a long-pass filter, to eliminate stray reflections of the laser beam, and through a 10-nm-bandwidth spike filter.

System	Wavelength λ(Å)	Excited Product	Energy Denominator $\Delta\omega$(cm^{-1})	Line Width (MHz)	$\mu_1{}^2$ (debye)^2	$\mu_2{}^2$ (debye)^2	σ/I (cm^4/watt)	Remarks
H$_2$	1930	E,F $^1\Sigma_g{}^+$(e) (ν=2, j=2)	45 x 10^3	10^6(a)	1(b)	1	1.8 x 10^{-31}	B $^1\Sigma_u{}^+$ intermediate state
CO	1930	F $^1\Sigma^+$	$\underline{13 \times 10^3}$(A$^1\Pi$) 10 (a$^3\Pi$)	10^6(a)	1 10^{-5}	$\underline{0.01}$ ~10^{-3}	1.6 x 10^{-32}	A$^1\Pi$ and a$^3\Pi$ intermediate states (d)
N$_2$	1930	E $^3\Sigma_g{}^+$	50 x 10^3	10^6(a)	1	0.01	1.1 x 10^{-33}	b' $^1\Sigma_u{}^+$ intermediate state
Kr	1930	Kr* 4p^{5}6p (103,762cm^{-1})	30 x 10^3	10^6(a)	1	1	3 x 10^{-31}	4p^{5}5s intermediate state
I$_2$	1930(e)	0442 $^1\Sigma_g{}^+$ (^{1}S+^{1}S)	4 x 10^3	10^6(a)	1	0.1	1.7 x 10^{-30}	D $^1\Sigma_u{}^+$(f) intermediate state
Xe	2484	Xe* (5p^{5}6p)	28 x 10^3	10^6(a)	1	1	5 x 10^{-32}	5p^{5}6s intermediate state
	1930	Xe+	16 x 10^3	~10^9(g)	1	1	1 x 10^{-34}	
N$_2$O	2484	N$_2$ + O*	28 x 10^3	~10^9(g)	~1	~10^{-3}	~3 x 10^{-35}	weak intermediate state amplitude

(a) Estimated linewidth. (b) J. E. Hesser, J. Chem. Phys. 48, 2518 (1968). (c) H. M. Crosswhite, The Hydrogen Molecule Wavelength Tables of Gerhard Heinrich Dieke (Wiley-Interscience, N.Y.). (d) P. H. Krupenie, The Band Spectrum of Carbon Monoxide, NSRDS-NBS No. 5 (USGPO, Washington, D. C., 1966). (e) R. S. Mulliken, J. Chem. Phys. 55, 288 (1971). (f) J. A. Myer and J. A. Samson, J. Chem. Phys. 52, 716 (1970).

172

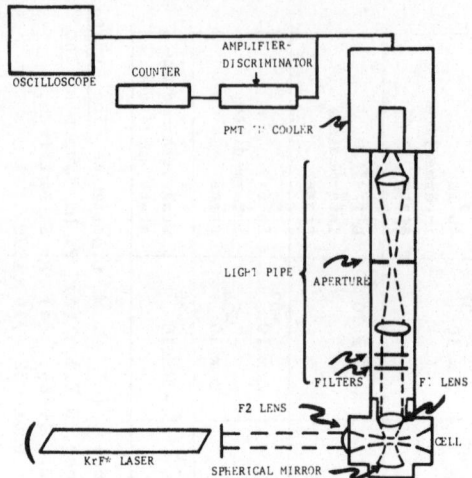

Fig.4 Illustration of the experimental apparatus indicating the laser source, the sample cell, and the photon counting detection system.

A thermo-electrically-cooled RCA C31034 photomultiplier tube was used to measure the fluorescence. The PMT output could be monitored in either analogue (directly on a Tektronix 585 oscilloscope) or digitally (with a PAR 1120 amplifier-discriminator and a Systron-Donner 6151 counter) mode.

In these experiments, the cell was filled with a mixture of N_2O in Xe at total pressures from 0-760 torr. The Xe:N_2O was pre-mixed in a sample cylinder in concentrations from 6000:1 to 500:1. Upon irradiation by the KrF* laser, the sample emitted a strong fluorescence signal in the range 520-570 nm. This signal was not present in the spectrum of pure Xe. As shown in Fig. 5,

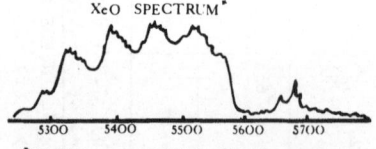

*Ref: J.R. Murray and C.K.Rhodes, J.Appl. Phys. 47, 5043(1976)

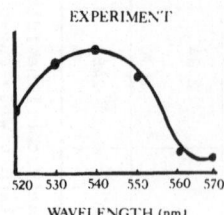

Fig.5 Xenon oxide and experimentally observed fluorescence spectra: The upper spectrum is a partially resolved XeO spectrum illustrating the collisionally induced auroral transition; the lower spectrum is that observed in the present experiments.

the fluorescence spectrum that we observed in Xe:N$_2$O emission corresponds, within the limits of our detection apparatus (10 nm resolution), to XeO* excimer emission [16].

In order to establish that two-photon absorption was involved in the formation of the radiating species, the dependence of fluorescence intensity on laser power was measured. Attenuators of 70%, 50%, and 35% transmittance were used to vary the ultraviolet input to the gas sample. As shown in Fig. 6,

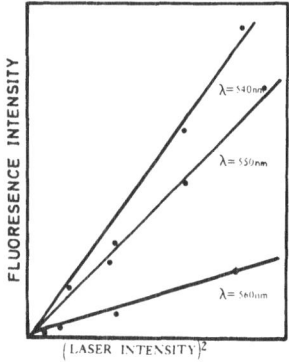

Fig.6 Fluorescence intensity of the observed signal illustrated as a function of the square of the laser intensity (248.4 nm). Curves for various fluorescent wavelengths are shown.

the fluorescence intensity in the spectral region of interest varied as the square of laser intensity, a behavior characteristic of two-photon absorption.

The results of these preliminary measurements are found to be in reasonable accord [14] with the estimates given for Xe and N$_2$O in Table I. Therefore, we conclude that two-photon amplitudes of this nature can be observed generally in wide classes of materials. Mechanisms of this type should (1) enable the study of many highly excited molecular states whose selective excitation would be impossible by more conventional means and (2) provide an important tool for kinetic analysis of highly excited electronic states.

4. Concluding Comments

Multiquantum excitation processes are used routinely for high resolution spectroscopic purposes in the visible spectral region and in the infrared for the performance of isotopically selective molecular dissociation, frequency conversion, and molecular spectroscopic analyses. A particular example, involving optical down-conversion, has been examined to illustrate the manner in which the electromagnetic degrees of freedom can manifest themselves in the dynamics of nonlinear amplitudes.

Recent technological developments in ultraviolet sources enable the observation of new classes of nonlinear phenomena in the ultraviolet region, a fact which parallels the previous achievements in the infrared. Estimates of two-quantum amplitudes have been given for several atomic and molecular materials and experiments confirming these values have been conducted in

Xe/N_2O mixtures at 248.4 nm. In this case the primary mechanism of absorption is two-photon excitation of xenon atoms which subsequently produce O(^{1}S) species by electronic energy transfer. We anticipate that these methods will find wide applicability in the near future for the study of the dynamics of electronically excited atomic and molecular materials.

5. Acknowledgments

The author gratefully acknowledges fruitful conversations with several individuals including W. K. Bischel, J. Bokor, D. Kligler, A. Lau, D. Pritchard, and H. Pummer as well as the expert technical assistance of W. Birkley.

References

1. P. Franken, A. Hill, C. Peters, and G. Weinreich, Phys. Rev. Lett. 7, 118 (1961)

2. J. Armstrong, N. Bloembergen, J. Ducing, and P. Pershan, Phys. Rev. 127, 1918 (1962); D. A. Kleinman in Laser Handbook, Vol. 2, ed. F. T. Arecchi and E. O. Schulz-DuBois (North-Holland Publishing Co., Amsterdam, 1972) p. 1229

3. P. L. Kelley, Phys. Rev. Lett. 15, 1005 (1965); R. Y. Chiao, E. Garmire, and C. Townes, Phys. Rev. Lett. 13, 479 (1964); S. A. Akhmanov, R. V. Khokhlov, and A. P. Sukhorukov in Laser Handbook, Vol. 2, ed. F. T. Arecchi and E. O. Schulz-DuBois (North-Holland Publishing Co., Amsterdam, 1972) p. 1151

4. I. D. Abella, N. A. Kurnit, and S. R. Hartmann, Phys. Rev. 141, 391 (1966); S. L. McCall and E. L. Hahn, Phys. Rev. Lett. 18, 908 (1967); E. Courtens, Phys. Rev. Lett. 21, 3 (1968)

5. D. E. Pritchard, J. Apt, T. W. Ducas, Phys. Rev. Lett. 32, 641 (1976); M. D. Levenson and N. Bloembergen, Phys. Rev. Lett. 32, 645 (1974); T. W. Hänsch, K. Harvey, G. Meisel, and A. L. Schawlow, Opt. Commun. 11, 50 (1974); G. Grynberg, F. Biraben, M. Bassini, and B. Cagnac, Phys. Rev. Lett. 37, 283 (1976); J. E. Bjorkholm and P. F. Liao, Phys. Rev. A14, 751 (1976)

6. R. V. Ambartzumian, V. S. Letokhov, E. A. Ryabov, and W. V. Chekalin, ZhETF Pis. Red. 20, 597 (1974) [JETP Lett. 20, 773 (1974)]; J. L. Lyman, R. J. Jensen, J. Rink, C. P. Robinson, and S. D. Rockwood, Appl. Phys. Lett. 27, 87 (1975)

7. W. E. Barch, H. R. Fetterman, and H. R. Schlossberg, Opt. Commun. 15, 358 (1975)

8. R. R. Jacobs, D. Prosnitz, W. K. Bischel, and C. K. Rhodes, Appl. Phys. Lett. 29, 710 (1976)

9. W. K. Bischel, P. J. Kelly, and C. K. Rhodes, Phys. Rev. A13, 1817 (1976)

10. H. Pummer, W. K. Bischel, and C. K. Rhodes, to be published

11. W. K. Bischel, P. J. Kelly, and C. K. Rhodes, Phys. Rev. A13, 1829 (1976)

12. A. M. Bonch-Bruevich and V. A. Khodovoi, Usp. Fiz. Nauk. 93, 71 (1967) [Sov. Phys.--Usp. 10, 637 (1968)]; P. F. Liao and J. E. Bjorkholm, Phys. Rev. Lett. 34, 1 (1975); P. F. Liao and J. E. Bjorkholm, Opt. Commun. 16, 392 (1976)

13. P. L. Kelley, H. Kildal, and H. R. Schlossberg, Chem. Phys. Lett. 27, 62 (1974)

14. D. Kligler, D. Pritchard, and C. K. Rhodes, private communication

15. R. Burnham and N. Djeu, Appl. Phys. Lett. 29, 707 (1976)

16. J. R. Murray and C. K. Rhodes, J. Appl. Phys. 47, 5043 (1976); C. D. Cooper, G. C. Cobb, and E. L. Tolnas, J. Mol. Spectros. 7, 223 (1961)

Raman Emission at 285 nm from Sn Vapor

N. Djeu

Laser Physics Branch, Naval Research Laboratory
Washington, DC 20375, USA

Highly efficient wavelength conversion of the KrF laser output at 248 nm
to 285 nm has been demonstrated via the resonant Raman process in Sn vapor.
This new UV radiation source is potentially useful in the photodissociation
isotope separation of UF_6.

The relevant energy levels in Sn responsible for the Raman process are
shown in Fig.1. The pump photon is nearly resonant with the (5 p^2 $^3P_2$ -
5d $^3F_2$) transition at 248.3 nm, creating a large Raman gain at the coupled
(5d $^3F_2$ - 5p^2 $^1D_2$) transition. Investigations of the frequency dependence
of the generated output on pump frequency showed that the stimulated emission arises almost entirely from the Raman process, and not the two-step
single-photon process. Thus, photon conversion efficiencies as high as
100% are in principle possible for arbitrarily long pump pulses.

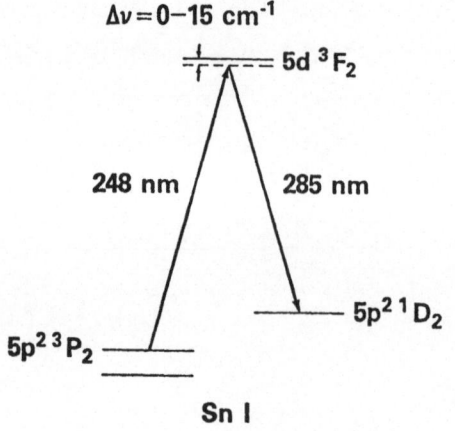

$\Delta\nu = 0{-}15$ cm^{-1}

5d $^3F_2$

248 nm

285 nm

5p^2 $^1D_2$

5p^2 $^3P_2$

Sn I

Fig.1 Energy levels in Sn used for
the generation of Raman emission

A diffraction limited, tunable KrF laser output served as the pump in
the experiment. The 8 nsec, 10 μJ laser pulse was focussed into a column
of Sn vapor to give a peak power density of about 100 MW cm^{-2}. The Sn vapor
was produced in an alumina tube of 2 cm diameter with a 10 cm long molybde-
num wick at the center. The tube was heated in a vacuum by a 5 cm long
tungsten coil and was additionally insulated in the central region by a

zirconia cylinder. With this arrangement it was possible to reach a temperature of 2000°K to give a Sn vapor pressure of 3 Torr. A fraction of the beam emerging from the heat pipe was deflected by a quartz flat into a 1 m monochromator with OMA detection for spectral analysis. The remainder of the beam was split by a prism into its 248 nm and 285 nm components which were then detected simultaneously by two photodiodes.

Stimulated emission near 285.1 nm was observed for detunings of up to 0.1 nm on either side of the $(5p^2\ ^3P_2 - 5d\ ^3F_2)$ resonance. The bandwidth of the generated output was approximately 0.05 nm, the same as that of the KrF pump. For pump frequencies near resonance, the contributions of Raman and two-step single-photon processes to emission at the coupled transition are determined by the relative rates of relaxation of the levels involved [1]. In the present system, since the relaxation rate of the initial state is much smaller than that for the intermediate state, one would expect the Raman contribution to predominate. This was indeed found to be the case, as the frequency of the generated output was observed to tune in unison with that of the pump. The Raman origin of the stimulated emission is important inasmuch as it permits the generation of output in the absence of an inversion between the intermediate and final states.

Measurements of the time resolved pulses at 248 nm and 285 nm with calibrated photodiodes showed a peak power conversion efficiency of 40% when the KrF laser was tuned to exact resonance with the pump transition. The Raman pulse was terminated a little more than halfway through the pump pulse. Rough calculations showed that medium saturation, i.e., the depletion of initial state atoms in some volume of interaction, was probably the cause of the premature ending. It is almost certain that higher conversion efficiencies will result from focussing the pump beam more tightly to give a higher pump intensity and/or using a longer Sn heat pipe. Furthermore, these results seem to be readily scalable to the conversion of much higher energy (~ 1J) KrF laser pulses.

Reference

1. I. M. Beterov and V. P. Chebotaev, *Progress in Quantum Electronics*, Vol. 3 (Part 1), Pergamon Press, Oxford, 1974.

Down Conversion of 351 nm Radiation for Fusion Lasers

J. Wilson and D. Ehrlich

Laboratory for Laser Energetics, University of Rochester
Rochester, NY 14627, USA

The properties required in a laser medium to make it suitable for a fusion laser are fairly well known[1,2]. Briefly, they can be summarized as:

(1)	Scaleability to a pulse energy of 1 Megajoule	
(2)	Average Power	10-100 Megawatts
(3)	Low Gain	$g_o \sim 10^{-2}$ cm^{-1}
(4)	Efficiency	$> 10\%$
(5)	Laser Pulse duration	~ 1 nanosecond
(6)	Wavelength	visible

It is a formidable task to provide all these properties in a single medium. It seems natural then to ask if some of the properties could be provided by one medium, the remainder by another. Expressed differently, could one take a laser that can be scaled efficiently to the energy required, but whose wavelength and pulse length are incorrect, and then convert the wavelength and pulse length in a second medium? The primary laser could be an excimer laser or a carbon dioxide laser. In this paper however, the emphasis is on excimer lasers, in particular XeF. It will be assumed that the primary laser can be scaled to the energies required. Vapour phase dye lasers will be proposed as the conversion medium. The conversion medium should convert the XeF pump pulse from 354 nm to the visible, and from a pulse duration of about 100 nsecs to 1 nsec.

In order to store the energy for at least as long as the pump pulse duration, the conversion medium must have a relatively long radiative lifetime. The medium must also permit efficient energy extraction, and have low gain. The bandwidth must be of the order of 10^9 hertz in order to generate a 1 nsec pulse. This has led fusion workers to consider metastable atoms, particularly O, S, Se as laser candidates[1]. These atoms are generated in an excited state from a parent molecule (e.g. N_2O in the case of the O atom) by photolysis. There are difficulties with this approach due to deactivation of the excited state by the parent molecule, and also in regeneration of the parent molecule.

Dye lasers are not commonly considered as low gain media! However there are dye-like molecules with long enough radiative lifetimes to have low gain. Typical fluorescence bandwidths are around 10^{13} hertz, and so easily satisfy the requirement for short-pulse generation. Efficient down-conversion from the ultra-violet to the visible has already been demonstrated with dyes[3]. The major question is whether down-conversion can be performed efficiently in the nanosecond time-scale. In atoms this presents no difficulty since the energy is stored in a single level. In molecules the energy is stored in many levels, and efficient extraction will require rapid intralevel transfer rates. The transfer will have to occur on times shorter than the required pulse time. In this respect dyes may be more favorable than smaller molecules. For a large system with high beam quality, vapours are preferred over high index condensed phase media. Dye vapour lasers have already demonstrated advantages over comparable dyes in solution[4]. Thus vapour phase dye molecules may possess the conversion properties required.

A partial list of candidate organic molecules is given in Table 1 in order of increasing radiative lifetime.

Table 1. Some aromatic hydrocarbons which absorb at 354 nm.

COMPOUND	ϕ_F	τ_{RAD} (nsec)	λ_F (nm)	VAPOR PRESSURE	STRUCTURE
VAPOR DYES WHICH HAVE LASED					
POPOP	0.65	1.3	385	3 torr at 300°C	
αNPO	0.4	2.0	380		
BBO	0.88*	1.15*	380		
3, 6 DERIVATIVES OF PHTHALIMIDE		30	530		
OTHER ORGANIC COMPOUNDS					
ANTHRACENE	0.5	12*	400	53 torr at 226°C	
FLUORANTHENE	0.3*	50	460	30 torr at 217°C	
BENZO (ghi) PERYLENE		107*	420	B.P. at 500°C	
PYRENE	0.32*	400*	380	60 torr at 260 C	
BENZOPHENONE		8000	440		

*MEASURED IN SOLUTION

Conventional dyes (e.g. POPOP) have short radiative lifetimes in fluorescence, and consequently, high gain. What is required here is a molecule with a long fluorescence lifetime. In addition the quantum yield for fluorescence, ϕ_F should be of order unity, if the conversion is to be efficient. An estimate of the radiative lifetime required may be obtained from the gain and the energy stored per unit volume. Assuming that an energy storage density of at least 10^{-3} joule/cc. is required implies an upper laser level population of 10^{16} molecules/cc. For a small signal gain $g_0 = 10^{-2} cm^{-1}$, at 500 nm, with a 10 nm bandwidth, the radiative lifetime necessary is 10^{-5} secs. The only listed candidate with a sufficiently long lifetime is benzophenone. The lifetime of benzophenone has been measured in the vapour phase [5]. Benzophenone is a primary candidate for a fusion laser converter. However, benzophenone is actually just one example of a class of organic molecules with long singlet radiative lifetimes. Other examples are acetophenone, biacetyl, benzaldehyde and derivatives of these and similar molecules.

Experimentally the low gain of benzophenone and similar long lifetime molecules presents a difficulty. In order to generate an excited state population of 10^{16} molecules/cc, in a volume of 2 cc's, a minimum energy of 12 millijoules is required. This amount of pump energy was not available to us. Instead a Nd:Glass laser was available, which by doubling the output with a CDA crystal, then summing the doubled output with the balance of the input in a type II KDP crystal, could be made to generate about 0.5 mjoule of third harmonic at 351 nm. This wavelength is close enough to that of XeF that it should simulate it well for molecules with a broad absorption spectrum.

With the pump energy available it seemed wisest to start with fluoranthene, and work gradually towards longer radiative lifetimes. In [6], BERLMAN quotes the decay rate of fluoranthene as 50 nanosecs., and the radiative lifetime (obtained from the integrated absorption coefficient) as 15 nanosecs. Thus there seems to be some doubt about the actual decay time. Fluoranthene has a good shift between absorption and emission, with peak emission around 440 nm[6].

The experiments were performed with longitudinal pumping rather than the transverse pumping used previously with nitrogen laser pumps[7,8]. Nitrogen lasers have been unsuccessful in longitudinal pumping due to their highly divergent output, but the well-collimated beam available to us should be suitable for longitudinal pumping.

The apparatus consists of an oscillator, amplifier, conversion crystals and dye cell.

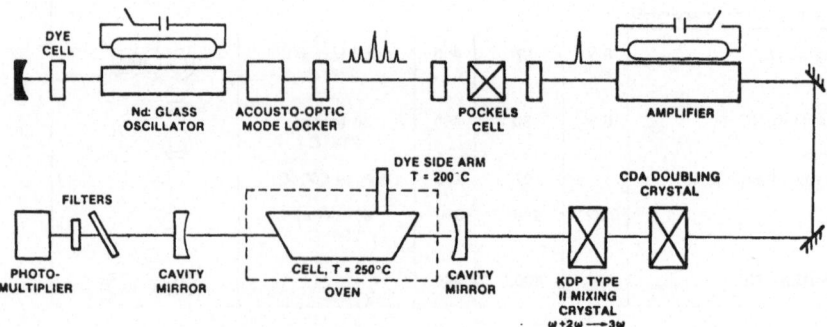

<u>Fig. 1.</u> Schematic diagram of the apparatus.

An acousto-optic/saturable dye mode-locked oscillator produces a train of mode-locked pulses. A single pulse, about 0.5 nsec in duration, is selected by a Pockels cell. This pulse is amplified and then converted to the third harmonic in the crystals. The third harmonic pulse is passed into a cell containing the dye vapour. The cell has Brewster's angle windows of quartz, and a path length of 20 cms. An oven heats the cell to about 250°c, and the vapour pressure of the dye is determined by a cold finger sticking out of the oven, and heated separately.

The upper level decay rate of fluoranthene was observed by monitoring the fluorescence of the fluoranthene with a fast photo-multiplier (RCA C70045) at various wavelengths selected by an interference filter in front of the photo-multiplier. The results are shown in fig. 2. The decay rate was found to be a function of the wavelength of the fluorescence, but not of the dye vapour pressure. Thus at short wavelengths, i.e. 420 nm, the decay time is 4.3 nanoseconds, but for longer wavelengths, 460 nm to 560 nm, the decay time is 21 nanoseconds.

Using the same technique we measured similar values in flouranthene solutions. The results suggest that there are at least two excited levels involved - one with a radiative lifetime of 4.3 nsecs, and one with a longer time, of the order of 21 nsecs. This may explain the discrepancy between the previous indirect measurements.

After measuring the fluorescence decay rate, we attempted to generate laser action by installing a cavity on the cell. The input mirror transmitted

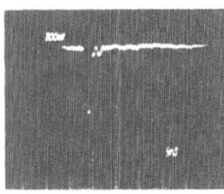 **A) LASER PULSE**

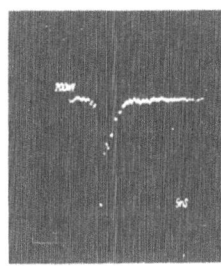 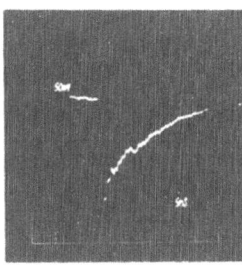 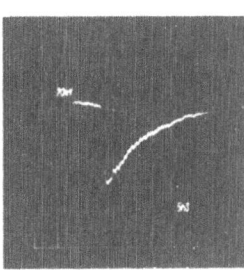

B) 420 nm   C) 460 nm   D) 560 nm

Fig.2. Oscillograms of fluorescence from fluoranthene at several wavelengths. A pump laser pulse is also shown for comparison.

the 351 nm pump light, but was 99.5% reflective at 440 nm. The output mirror had approximately 0.2% transmission at 440 nm. No laser action was observed. The decay time of fluoranthene is equivalent to 7 round trips in our cavity. This is probably too short for laser generation.

With the cavity installed, we also attempted to generate laser action in pyrene. Again no laser action was observed. This molecule was actually a bad choice due to its low absorption at 351 nm. The decay time of fluorescence was measured to be 70 nsecs., considerably shorter than in solution.

Summary
Organic molecules with long radiative lifetimes have been shown to be potential candidates for converting excimer lasers into fusion lasers. Benzophenone and similar molecules are the most promising candidates. Experiments on fluoranthene and pyrene have measured the vapour phase lifetime, but did not generate laser action.

References

1. J. R. Murray and C. K. Rhodes, J. Applied Physics, $\underline{47}$, 5041, (1976).
2. J. Wilson and D. O. Ham, Laser Focus, 12, 38, Nov. 1976.
3. P. P. Sorokin, J. R. Lankard, E. C. Hammond, and V. L. Moruzzi I.B.M. Journal of Research and Development, $\underline{11}$, 130 (1967).
4. L. G. Pikulik, V. A. Yakovenko, and A. D. Das'ko, Zhurnal Prikladnoi Spektroskopiee, $\underline{23}$, 493 (1975).
5. G. E. Busch, P. M. Rentzepis, and J. Jortner, Chem. Phys. Letts., $\underline{11}$, 437, (1971).
6. I. B. Berlman, Handbook of Fluorescence Spectra of Aromatic Molecules, Academic Press, New York, (1971).
7. B. Steyer and F. P. Schäfer, Applied Physics, $\underline{7}$, 113, (1975).
8. P. W. Smith, P. F. Liao, C. V. Shank, T. K. Gustafson, C. Lin, and P. J. Maloney, Applied Physics Letters, $\underline{25}$, 144, (1974).

Part V

Applications

Final State Energy Distributions for Exoergic Reactions

S. Fischer

Institut für Theoretische Physik, Technische Universität München
8046 Garching, FRG

1. Introduction

The rapid development of laser techniques makes it possible to get detailed information about product state distributions for reactive collisions. It is the objective of this paper to present a model which makes a priori predictions about vibrational, translational and rotational final state distributions. The vibrational and the translational degrees of freedom are treated quantum mechanically in close analogy to a collinear system. The rotation is treated statistically. The application of the saddle-point method brings into the theory a distribution parameter which might be looked at as the generalization of the scheme of an internal temperature to systems for which the a priori probability of final state population is still dictated by the dynamics. Applications to very exothermic triatomic exchange reactions are discussed and the theoretical predictions are compared with experiments. The work has been performed in collaboration with G. Venzl [1,2,3].

2. Collinear Triatomic Exchange Reactions [1]

For the collinear arrangement the transition matrix can be expressed as

$$T_{fi} = \iint dr_f dR_f \left\{ \Psi_{nl}(r_f,R_f) \left[V(r_i,r_f) - \bar{V}(R_f) - v_f(r_f) \right] \Psi_{ok}(r_f,R_f) \right\}. \tag{1}$$

$\Psi_{nl}(r_f,R_f)$ is the Born-Oppenheimer wave function for the product state, which is approximated by

$$\Psi_{nl}(r_f,R_f) \approx \frac{1}{\hbar^{1/2}} \delta(R_f - R_f^n(\varepsilon_\ell)) \; \phi_n^f(r_f) . \tag{2}$$

The δ-function is an approximation to the outgoing translational wave function, and $\phi_n^f(r_f)$ is the product vibrational wave function, here approximated by the Hermite function for the harmonic oscillator. The product of the effective interaction $V_{eff} = V(r_i,r_f) - \bar{V}(R_f) - v_f(r_f)$ and the incoming scattering wave function $\Psi_{ok}(r_f,R_f)$ in the vibrational ground state (m=0) is approximated by the product of a Gaussian in R_i and the reactant vibrational wave function $\phi_o^i(r_i)$

$$V_{eff} \; \Psi_{ok}^+(r_f,R_f) \approx \exp\left[- \frac{(R_i - R_i^*)^2}{2(\gamma \sigma r_f^o)^2} \right] \; \phi_o^i(r_i). \tag{3}$$

R_i^* is representative for the fall-off region of the incoming wave packet.

$$R_i^* = \bar{r}_i + \gamma_i \, r_f^*$$

where

$$r_f^* = r_f^o \, (2n^*+1)^{1/2} + \bar{r}_f \;.$$

Further definitions are: the equilibrium distance $\bar{r}_f$ and $\bar{r}_i$ for the final and initial diatomics, the zero-point amplitude r_f^o, the mass ratios

$$\gamma_i = \frac{m_1}{m_1+m_2} \;, \qquad \gamma_f = \frac{m_3}{m_2+m_3} \;,$$

and

$$\gamma = 1 - \gamma_i \gamma_f$$

where the attacking atom has the mass m_1. n^* is the vibrational quantum number of the product diatomic which corresponds to the energy released in the process which is denoted as E. The width σ finally is evaluated within the harmonic model as

$$\sigma = 0.83 \, (n^* + 1/2)^{-1/6} \, A_\perp^{-1/3} \, \gamma^{-2/3} \tag{4}$$

and $A_\perp$ is the atractiveness parameter specifying the location of the outgoing potential as

$$\bar{V}_f(R_f) = E(1-A_\perp) \, \exp\left\{ \frac{\gamma_i^2}{4} \left(\frac{r_f^o}{L}\right)^2 - \frac{R_f - \gamma_i \bar{r}_f - \bar{r}_i}{L} \right\} \tag{5}$$

with the interaction length L.

The main approximations concern the δ-function approximation for the outgling translational wave function and the Gaussian approximation for the combined effective interaction and the incoming wave function. The latter one is motivated by the fact that the effective potential is negligible small for $r_f < \bar{r}_f$ and $r_i \gtrless \bar{r}_i$ and the incoming wave function peaks at the fall-off region characterized by R_i^* or r_f^*. The width takes into account the fact that this peak penetrates deeper into the interaction region at $\bar{r}_i$, $\bar{r}_f$ if $A_\perp$ is small, that means if the fall-off takes place largely in the outgoing channel. The approximations have been tested against exact quantum calculations and they seem to work very well. It is now easy to evaluate the integral. The result is given in [1].

$$|T_n^v|^2 = \frac{2^{-n}}{n!} \, |\alpha|^n \, H_n(g)|^2 \, \exp\left\{ - \frac{(X(\epsilon_\ell) - \tilde{X})^2}{\tilde{\sigma}^2} \right\} \, \epsilon_\ell^{-1} \tag{6}$$

with

$$\alpha^2 = \frac{1 + \sigma^2(\tilde{\gamma}_i - 1)}{1 + \sigma^2(\tilde{\gamma}_i + 1)} \tag{7}$$

$$g \approx \frac{\frac{1}{\gamma} (2\overset{*}{n}+1)^{1/2} + \tilde{\gamma}_i^{1/2} \left[\sigma^2 - \frac{1}{\gamma} \left(\frac{\gamma_f}{\gamma_i}\right)^{1/2} \left(\frac{k_f}{k_i}\right)^{1/2} \tilde{X} \right]}{\left[(\overset{\vee}{\gamma}_i \sigma^2 + 1)^2 - \sigma^4 \right]^{1/2}} \tag{8}$$

and

$$\tilde{\gamma}_i = \gamma_i (\gamma_i \gamma_f)^{1/2} \left(\frac{k_i}{k_f}\right)^{1/2} . \tag{9}$$

k_i and k_f are force constants, H_n is the Hermite polynomial, $X(\varepsilon_\ell)$ is the classical turning point to the translational energy ε_ℓ

$$X(\varepsilon_\ell) = \frac{L}{r_i^0} \ln \frac{(1-A_\perp) E}{\varepsilon_\ell} \tag{10}$$

and $\tilde{X}$ is a shift caused by the dynamics

$$\tilde{X} = \frac{\tilde{\gamma}_i^{1/2}}{\gamma} (2\overset{*}{n}+1)^{1/2} \frac{1 + \frac{1}{\gamma}(\frac{\gamma_f}{\gamma_i})^{1/2} (\frac{k_f}{k_i})^{1/2} (1+\tilde{\gamma}_i)}{1+\sigma^2 + \frac{\tilde{\gamma}_i}{\gamma^2} \frac{\gamma_f k_f}{\gamma_i k_i} (1+\overset{\sim}{\gamma}_i) + 2 \frac{\gamma_i \gamma_f}{\gamma}} . \tag{11}$$

$\tilde{\sigma}$ finally is the effective width for the translational distribution

$$\tilde{\sigma}^2 = \frac{1 + \sigma^2(1+\tilde{\gamma}_i)}{1 + \sigma^2 + \frac{\tilde{\gamma}_i}{\gamma^2} \frac{\gamma_f k_f}{\gamma_i k_i} (1+\tilde{\gamma}_i) + 2 \frac{\gamma_i \gamma_f}{\gamma}} . \tag{12}$$

In the expression for g $X(\varepsilon_\ell)$ is replaced by $\tilde{X}$ in order to get the factorisation. Thus we find a Hermite polynomial which is for typical values of g similar to a Poisson distribution for the vibrational final-state population and a Gaussian-type function for the translational distribution. The Gaussian is a function of ε_ℓ slightly asymmetric with a longer tail towards larger translational energies, since the spacing of $X(\varepsilon_\ell)$ is not linear in ε_ℓ.

As test of the simplifying assumptions which enter in the collinear model the final state vibrational distribution was calculated for the series Mu, H, D, T +F$_2$ → Mu F, H F, D F, T F, + F. For this series of light-heavy-heavy reactions exact quantum mechanical calculations are available /4/. The results of the model calculations are shown in Fig.1. They agree within two percent with the exact calculations. It can be seen that the width of the distribution decreases with increasing number of accessible states n^* as $(n^*)^{-1/2}$ and the location of the maximum is shifted to higher values of n. As a second test the variation of the relative product vibrational energy $\langle f_n \rangle$ is

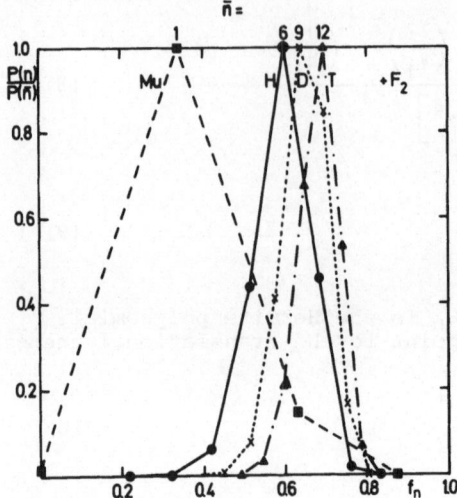

Fig.1 $X + F_2$-reaction for $X = Mu, H, D, T$

calculated as function of the attractiveness parameter. The almost linear increase for the LHH reaction and the small variation for HLL reactions is consistent with predictions by KUNTZ et al. [5].

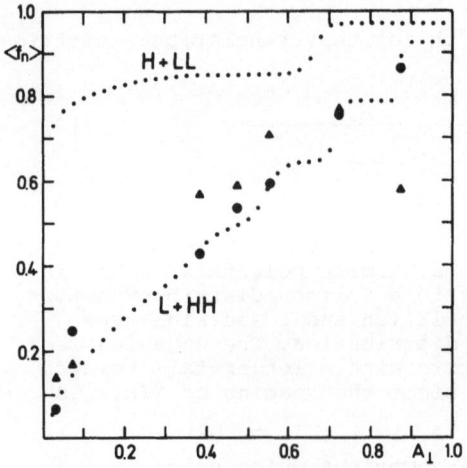

Fig.2 Dependence of vibrational energy $<f_n>$ on $A_\perp$; trajectory calculations, model calculations for a series of surfaces given by KUNTZ et al [5]

3. The Distribution Parameter and Discussion of the Results

The transition matrix (1) factorizes in a term which depends only on the vibrational quantum numbers, and one which depends only on the translational energy. We choose these factors as weighting factors for the corresponding partition function and add the rotational partition function for rotation around a fixed plane. The vibrational partition function reads:

$$F^V = \sum_n \frac{2^{-n}}{n!} |H_n(g)|^2 e^{-\beta\omega\hbar} = \left[1 - |\alpha|^2 e^{-\beta\omega}\right]^{-1/2} \exp\frac{2|\alpha g^2|e^{-\beta\omega}}{1+ \frac{\alpha^2}{|\alpha|}e^{-\beta\omega}} \quad (13)$$

and as translational function

$$F^{tr} = - \int d\varepsilon_\ell \exp \{- \frac{1}{\tilde{\sigma}^2} (X(\varepsilon_\ell) - \tilde{X})^2 - \beta\varepsilon_\ell \} \qquad (14)$$

and finally as rotational function

$$F^{rot} = \int dJ e^{-\beta\omega_J} \qquad . \qquad (15)$$

The distribution parameter β is determined by means of the energy balance equation

$$E = \omega \left\{ \frac{2|\alpha g^2| \exp(-\beta\omega)}{(1+ \frac{\alpha^2}{|\alpha|} \exp(-\beta\omega))^2} + \frac{|\alpha|^2 \exp(-2\beta\omega)}{1 - |\alpha|^2 \exp(-2\beta\omega)} \right\} + \varepsilon_{tr} \quad (16)$$

$$+ \frac{1}{2}\beta^{-1} + \sum_j \omega_j g_j^2 \exp(-\beta\omega_j)$$

where the translational energy is given by

$$\varepsilon_{tr} = \frac{\int d\varepsilon \, \varepsilon \exp \{- \frac{1}{\tilde{\sigma}^2}(X(\varepsilon) - \tilde{X})^2 - \beta\varepsilon\}}{\int d\varepsilon \exp \{- \frac{1}{\tilde{\sigma}^2}(X(\varepsilon) - \tilde{X})^2 - \beta\varepsilon \}} \approx \varepsilon^* \qquad . \qquad (17)$$

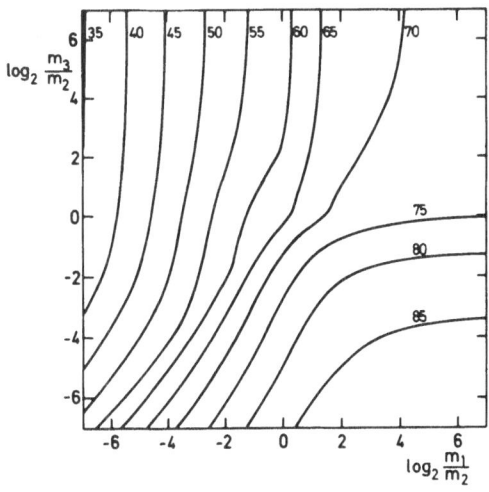

Fig.3 Average vibrational excitation $\langle f_V \rangle$ as a function of mass combinations. We used the H + Cl$_2$ potential parameters of Table 1 and chose the lightest mass to be 1 amu

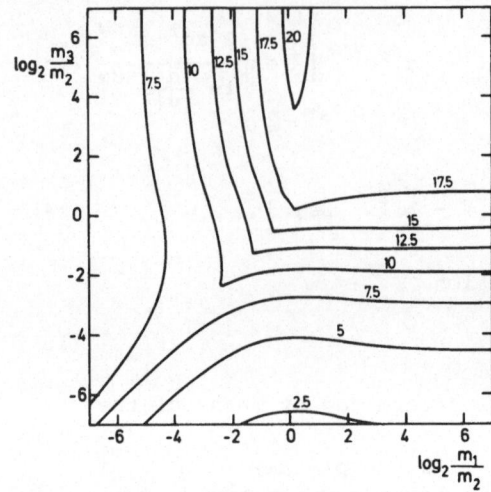

Fig.4 Average rotational excitation $\langle f_R \rangle$ as a function of mass combinations. Parameters as in Fig.3

In Figs.3 and 4 are drawn contour maps for the vibrational and rotational energy as a function of $\log_2 m_3/m_2$ and $\log_2 m_1/m_2$. The vibrational energy is smallest for the LHH mass combination ($m_1/m_2 \ll 1$ and $m_3/m_2 \approx 1$) and increases as m_1/m_2 increases. For $m_3/m_2 > 1$ this is mainly due to a decrease of γ or the scewing angle of the reaction path in a mass-weighted coordinate system. Smaller values of γ cause larger values of $\tilde{X}$ which yield less translational energy and thus more internal energy. Both, vibrational and rotational energy (Fig.4) are increasing. For $m_3/m_2 < 1$ the vibrational energy increases with increasing m_1/m_2 because the vibrational frequency ω of the product diatomic becomes smaller. This tends to increase the accessible number of quanta n^* and thus $\tilde{X}$, and also the coupling constant $|\alpha g^2|$. The variation of the vibrational energy in the lower right of the plot is due to the fact that the rotational energy is strongly dependent upon m_3/m_2 in that region (compare the following discussion and Fig.4), whereas the translational energy is practically constant.

Regarding the rotational energy, two main trends are found. The first one is the increase of rotational energy for increasing values of m_3/m_2 in the lower right of Fig.4. It is due to the reduction of vibrational frequency ω: Since the product $\beta\omega$ occurs in the exponential $\exp(-\beta\omega)$, a decrease of ω has in a zeroeth approximation to be compensated by an increase of β. The rotational energy which is within this model simply $1/2\beta$ is therefore reduced. The second trend is the small value of rotational excitation for the LHH mass combination and its increase with increasing m_1/m_2 until m_1 and m_2 are about equal. Interestingly these features come out of the model even though angular momentum conservation has not been considered explicitly.

In Fig.5 the translational energy distribution is shown. It is of the Gaussian type with a somewhat longer tail towards the

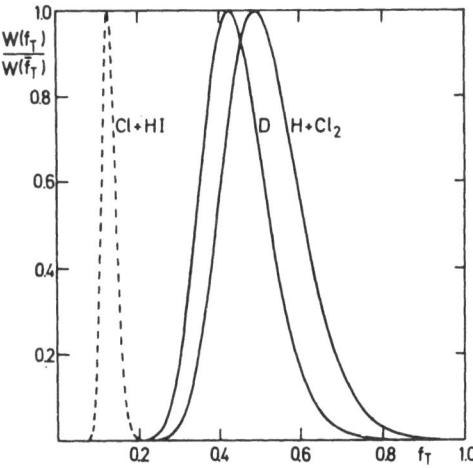

Fig.5 Translational energy distribution for Cl + HI and H,D + Cl_2. Potential parameters of Table 1

high translational energy side. This asymmetry would come out more pronounced for a potential with a steeper repulsive part as for instance the Lenard Jones potential. The width is very small for the **LHH** mass combination. In the **LHH** limit it increases, as m_l decreases. At the same time the maximum shifts towards higher translational energies. Such predictions could never come out of models in which the rotational and translational degrees of freedom are both treated statistically.

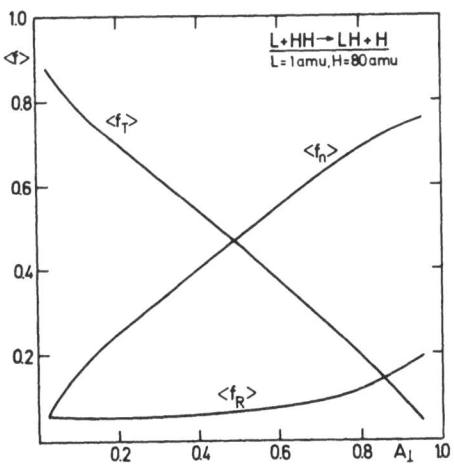

Fig.6 Energy disposal in a **LHH** reaction: average vibrational, rotational and translational energy as function of the attractiveness parameter A . The remaining potential parameters are from the H+Cl_2 surface (cf. Table 1)

In Fig.6 the vibrational energy $<f_n>$, the translational energy $<f_T>$, and the rotational energy $<f_R>$ are plotted as a function of the attractiveness parameter $A_\perp$ for the **LHH** mass combination. $<f_T>$ falls off almost linearly with increasing $A_\perp$, and $<f_n>$ increases correspondingly. Interestingly, $<f_R>$ stays about constant from small values of $A_\perp$ up to $A_\perp \approx 0.7$ and increases then remarkedly. This trend is also predicted by trajectory calculations.

Finally applications to some **HLH** and **LHH** reactions are shown in order to compare it with experimental results. The potential parameters used have been taken from the literature and are given in Table 1. The calculations were done restricting the sum in (13) to the energetically accessible values by imposing $\varepsilon_n \leq E$. This improvement only slightly changes the results if quantum states with $\varepsilon_n > E$ could contribute to the canonical vibrational partition function F^V, as it is the case for Cl+HI and Cl+DI (see Fig.7).

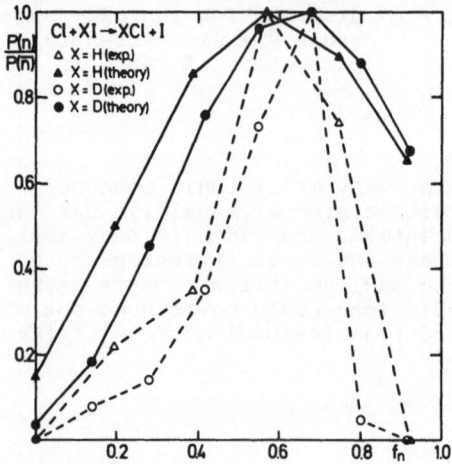

Fig.7 Vibrational distribution for the Cl+HI and Cl+DI reactions: the theoretical curves are obtained by the $g = g(\tilde{X})$ method (see text).

Table 1 Potential parameters for some triatomic exchange reactions A + BC → AB + C

	H+F$_2$	H+Cl$_2$	Cl+HI
$D_{BC}/Kcal \cdot mol^{-1}$	37.6	58.14	73.8
$\beta_{BC}/\text{Å}^{-1}$	2.92	2.008	1.776
$D_{AB}/kcal \cdot mol^{-1}$	141.1	106.86	106.4
$\beta_{AB}/\text{Å}^{-1}$	2.219	1.869	1.5
$E_c/kcal \cdot mol^{-1}$	2.36	2.42	0.11
$A_\perp$	0.41	0.381	0.24

Table 2 gives the average vibrational and rotational excitation of the products according to experiments and our
theory. The factorisation method of the collinear transition
matrix into a vibrational and a translational part which was
outlined in section 3 replaces $g(X(\varepsilon))$ by $g(X(\varepsilon^*)) \approx g(\tilde{X})$ and is
best justified if the translational distribution is sharply
peaked at ε^*. This method is therefore adequate for HLH reactions. It is used in the first column of theoretical results
in Table 2. Besides this, a second factorisation method is used

Table 2   Fraction of vibration and rotation for some triatomic
exchange reactions. Comparison of our theory with experimental
results

	Experiment:		Theory: $g = g(\tilde{X})$		Theory: $g = g(X_n)$	
	$\langle f_V \rangle$	$\langle f_R \rangle$	$\langle f_V \rangle$	$\langle f_R \rangle$	$\langle f_V \rangle$	$\langle f_R \rangle$
$H + F_2$	0.58	0.03	0.58	0.05	0.53	0.08
$D + F_2$	–	–	0.66	0.04	0.59	0.09
$H + Cl_2$	0.38	0.09	0.44	0.07	0.40	0.10
$D + Cl_2$	0.39	0.10	0.52	0.06	0.47	0.09
$Cl + HI$	0.70	0.13	0.58	0.29	–	–
$Cl + DI$	0.70	0.13	0.62	0.25	–	–

which replaces $g(X(\varepsilon))$ by $g(X(E-\varepsilon_n)) =: g(X_n)$. This method is
expected to be more accurate if the translational distribution
is less narrow and the rotational excitation is small. In the
case of LHH reactions, it yields a narrower vibrational distribution than the $g = g(\tilde{X})$ method and therefore compares better
with the experimental distributions which are even narrower
(Figs.8,9).

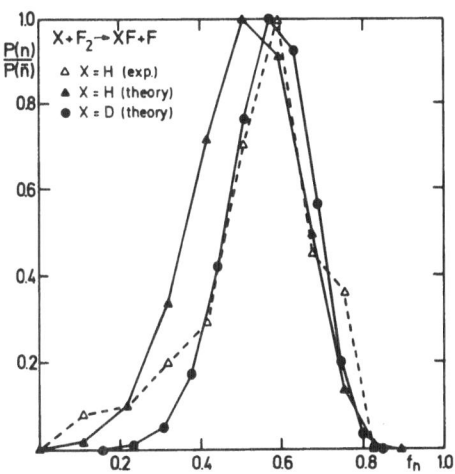

Fig.8 Vibrational di-
stributions for the H,D+F_2
reaction: the theoretical
curves are obtained by the
$g = g(X_n)$ method (see text).

194

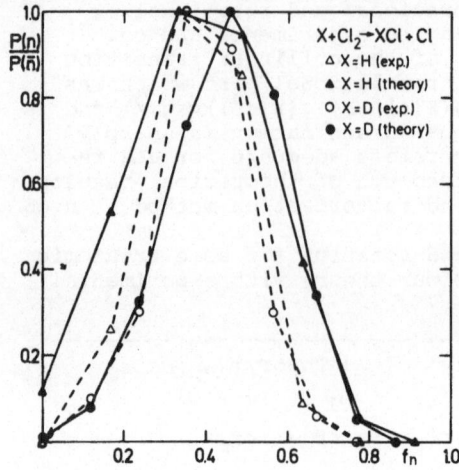

Fig.9 Vibrational distributions for the H,D + Cl_2 reactions: the theoretical curves are obtained as in Fig.8

Summarizing Figs.7,8 and 9, the theory describes quite well the average vibrational excitation for these reactions, whereas the widths of the distributions are always too large. This is partly due to the canonical analysis and partly to the statistical assumption that the whole phase space for rotations is equally probable. Furthermore, according to Table 2 a small amount of rotational excitation is predicted for LHH reactions, which is also found experimentally.

References

1. S.Fischer, G.Venzl: J.Chem.Phys., in press
2. S.Fischer, G.Venzl: subm. for publication in Chem.Phys.
3. S.F.Fischer: Ber.Bunsenges. 81, 197 (1977)
4. J.N.L.Connor, W.Jakubetz, J.Manz: Chem.Phys.Lett. 39, 75 (1976); Abstr. of the 4th Conf. on Molecular Beams, Amsterdam 1977
5. P.J.Kuntz, E.M.Nemeth, J.C.Polanyi, S.F.Rosner, C.E.Young: J.Chem.Phys. 44, 1168 (1966)

Reactions of Atoms with Vibrationally Excited Molecules

M. Kneba, K.J. Schmatjko, and J. Wolfrum

Abteilung Reaktionskinetik, Max-Planck-Institut für Strömungsforschung
3400 Göttingen, FRG

1. Introduction

Reactions of free atoms in the gas phase play an important part
in the development of high power lasers. Investigations of microscopic details of atom reactions led to the discovery of the hydrogen halide and other chemical lasers. Molecules formed in exothermic reactions of atoms often exhibit a strong vibrational and
rotational nonequilibrium excitation, which can be converted into
stimulated infrared emission. More recently atom reactions which
form electronically excited products have become very popular in
connection with lasers operating on one or more transitions in
the ultraviolet and visible wavelength region. Beside electronically
excited molecules the excited atoms itself can also be used in
electronic transition lasers.

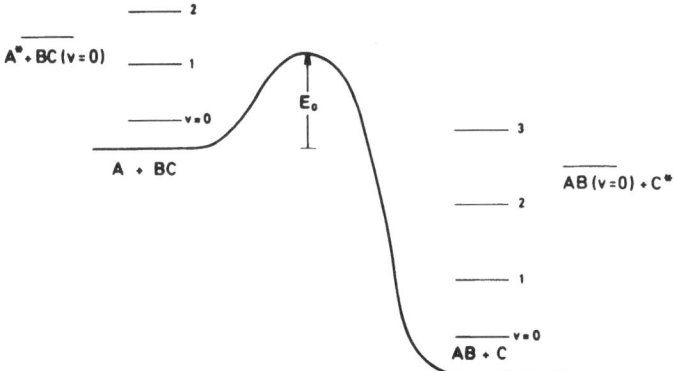

Fig. 1   Energy diagramm for reactions of atoms
with vibrationally excited molecules

As schematically shown in Fig. 1 two aspects of atom reactions in
connection with high power lasers and their application will be considered in the following paper :

(i) Microscopic details of the energy disposal resulting in population inversion of vibrational and electronic states.

(ii) Studies of the energy dependence of chemical reaction rates. Traditionally reaction rates are measured under conditions where the rate of reaction is slow compared to that of energy transfer. Simple atom reactions are interesting candidates for detailed studies of the effect of selective vibrational excitation by infrared lasers on chemical reactions in the gas phase [1].

2. Energy Distribution : The Reaction $O(^3P)$ + CN

The reaction of oxygen atoms with CN-radicals can proceed on two pathways

$$\Delta H_0^o \quad [kJ/mol]$$

(1a) $O(^3P) + CN(v) \longrightarrow CO(v') + N(^2D) \qquad -78$

(1b) $\longrightarrow CO(v') + N(^4S) \qquad -309$

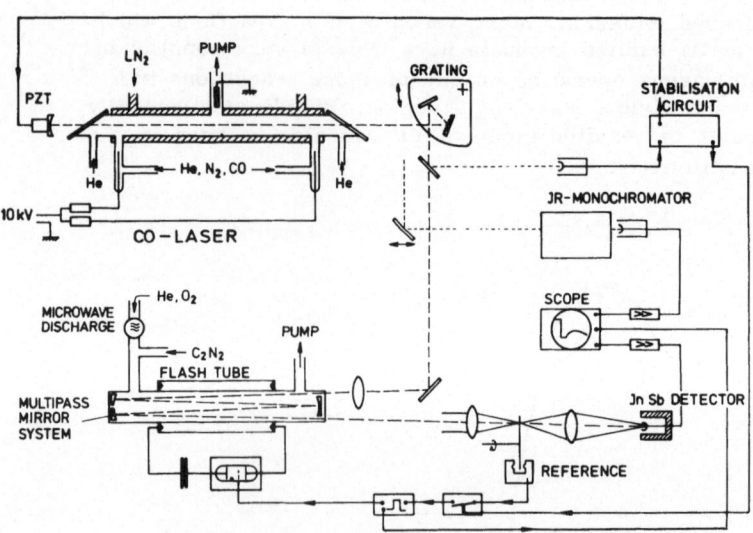

Fig. 2 Schematic diagram of the experimental apparatus for direct measurement of CO(v) from reaction (1) by infrared laser resonance absorption

On path (1a) the reaction energy is converted into energy of the metastable $N(^2D)$ atom, in path (1b) energy is available for population of high vibrational states in the CO molecules. In order to measure directly the energy distribution in the products formed in reaction (1), a combination of flash-photolysis and discharge-flow system is employed (s. Fig. 2). Using the high sensitivity of time - resolved resonance absorption with an infrared laser source low concentrations of product particles (down to 10^{10} cm^{-3}) could be monitored. This reduces the effect of energy exchange during the observation time as can be seen from the experimental curves in Fig. 3a.

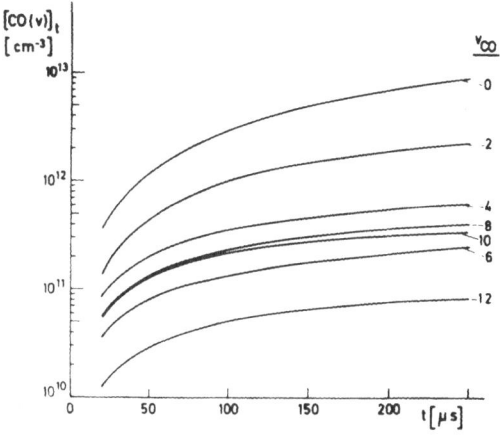

Fig. 3a
Measured concentration profiles of CO(v) formed in reaction (1)

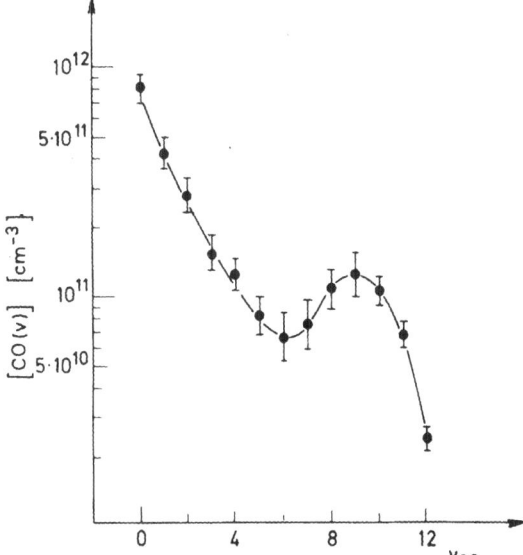

Fig. 3b
Absolute vibrational population of CO(v) measured at a reaction time of 40 μs

The form of the distribution measured at early times (see Fig. 3b) remains nearly unchanged as the reaction time proceeds. The two different parts of the observed vibrational energy distribution for

CO(v) can be understand by considering the dynamics of the two channels of the reaction. Three dimensional classical trajectory calculations [2] show on path (1a) a more statistical energy distribution in which the formation of CO (v = 0) predominates. This reaction path includes the electronic ground state of the NCO radical as intermediate reaction complex with a lifetime of about 10 vibrational periods. On path (1b) a more direct interaction on a repulsive surface occurs which channels the reaction energy mainly into vibrational excitation of the CO molecules. From the measured CO(v) distribution (Fig. 3) a high inversion of the nitrogen atoms produced in the reaction (1) is predicted. A direct observation of the electron state of the nitrogen atoms was made using the atomic resonance lines in the vacuum ultraviolet. Fig. 4 shows the population inversion by the time resolved absorption signals for the excited $N(^2S)$ and the ground $N(^4S)$ state. Unfortunately the extraction of this excitation energy by stimulated emission from the $N(2^2D_J) \longrightarrow N(2^4S_{3/2})$ transition at 520 nm is very difficult due to the low Einstein coefficient around 10^{-5} [3] of this emission which is prominent in the day glow [4]. However, other possibilities may arise in future research on lasers involving group V atoms.

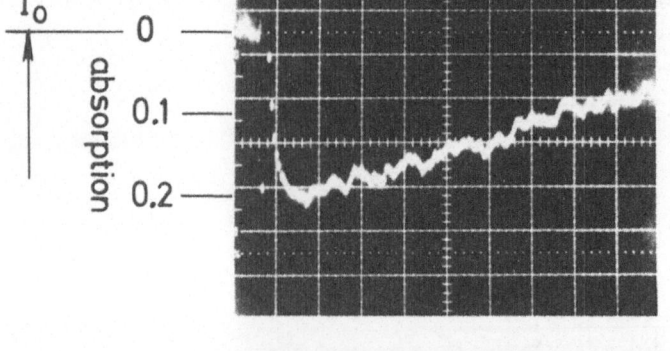

Fig. 4a

$N(^2D)$: 200 µs /div

Oscillogram showing the atomic line resonance absorption of $N(^2D)$ formed in reaction (1)

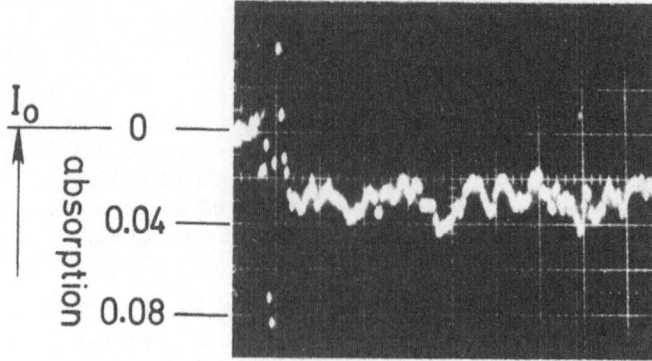

Fig. 4 b

$N(^4S)$: 100 µs / div

Oscillogram showing the atomic line resonance absorption of $N(^4S)$ formed in reaction (1)

3. Single Photon Chemistry in the Infrared : A + BC (v) Reactions

Laser light sources offer many new ways for a controlled transfer
of external energy into molecules. The high quantum flux, mono-
chromaticity, tunability, polarization and short pulse duration of
high power lasers now available in the infrared, visible and ultraviolet spectral range allow a very specific preparation of molecules in higher energy states. As an example the effect of selective vibrational excitation after single photon absorption from an
infrared laser on simple chemical reactions in the gas phase will
be described. As shown in Fig. 5 the reactions were chosen so
that vibrational excitation by the absorption of a single infrared
photon is comparable to the potential energy barrier of the reaction. The experimental arrangement must be able to distinguish

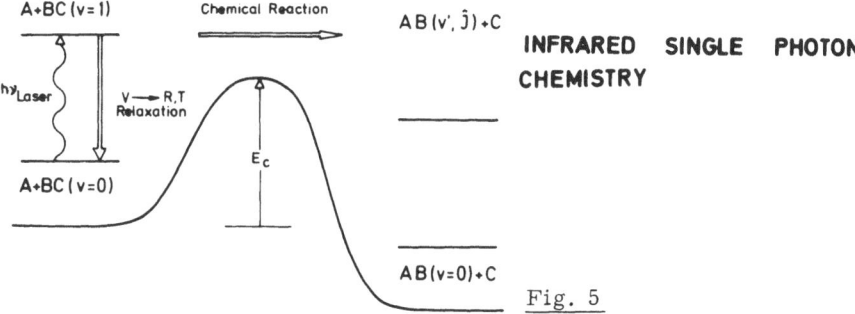

Fig. 5

between reactive and energy transfer processes removing the vibrationally excited molecules. This is achieved by determining the absolute concentration of the excited molecules by a measurement of
the relative population in the levels $v = 1$ and $v = 2$ after the rapid
equilibration of vibrational energy within the molecules using a
low degree of excitation. The concentration of the reacting atoms
is followed by time resolved chemiluminescence [5]. As shown in
Fig. 6 reaction products were analyzed by a molecular beam samp-
ling system and a quadrupole mass spectrometer.

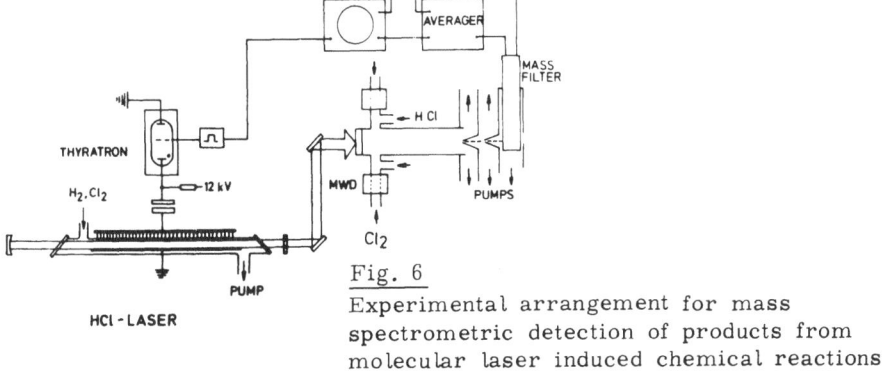

Fig. 6
Experimental arrangement for mass
spectrometric detection of products from
molecular laser induced chemical reactions

As first example the thermoneutral atom exchange reaction

(2) $Cl' + HCl \longrightarrow Cl'H + H$

will be considered. There exists a number of conflicting data on the existence or nonexistence of a substantial potential energy barrier in this reaction [6-9]. Isotopically selective vibrational excitation is a valuable tool to obtain new insights into the microscopic dynamics of such chemical reactions. With the experimental set up shown in Fig. 6 vibrational excitation of HCl occurs predominantly for $H^{35}Cl$ when using a chemical HCl laser.

$$H^n Cl + hv_L \longrightarrow H^{35}Cl (v = 1)$$

After laser excitation, formation of $H^{37}Cl$ can be observed by the time resolved mass signal (see Fig. 7). A signal is obtained during the transit time·of the volume excited by the laser in the flow reactor. From the absolute amount of $H^{37}Cl$ formed and the number of $H^{35}Cl(v=1)$ molecules produced by the laser pulse the rate of the reaction

(2') $^{37}Cl + H^{35}Cl(v=1) \longrightarrow H^{37}Cl(v=1,0) + {}^{35}Cl$

$k_2(v=1) = 10^{12,8} \ cm^3/mol \cdot s$ at 298K

can be determined. From the decay of $H^{37}Cl(v=0)$ in the presence of ^{35}Cl atoms the ground state reaction rate is obtained

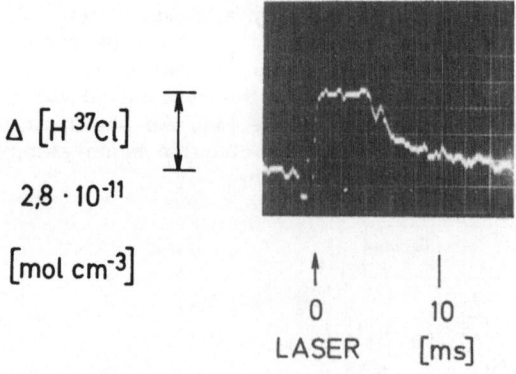

$\Delta [H^{37}Cl]$

$2,8 \cdot 10^{-11}$

$[mol \ cm^{-3}]$

0   10

LASER   [ms]

$^{37}Cl + H^{35}Cl (v=1) \longrightarrow H^{37}Cl (v=1,0) + {}^{35}Cl$

Fig. 7 Oscilloscope trace of $H^{37}Cl$ (mass peak m/e = 38) generated in reaction (2')

$$k_2 (v=0) = 10^{9,5} \text{ cm}^3 / \text{mol} \cdot \text{s} \qquad T = 298 \text{K}$$

Thus, the rate of reaction (2) increases by more than a factor of 10^3 due to the absorption of a single infrared photon. Our measurement of $k_2 (v=0)$ support the activation barrier found by Klein et al. [6]. The value for $k_2 (v=1)$ shows that reaction (2') makes a significant contribution to the fast removal of vibrational excited HCl in the presence of Cl atoms [5,10].

The effective vibrational relaxation by potentially reactive atoms is a very important limiting factor in high power chemical lasers. In the HCl laser chlorine as well as hydrogen atoms are present. We have studied the elementary steps in the $H + HCl(v)$ system also by using isotopic substitution. Fig. 8 shows the formation of DCl in the reaction (3'a)

(3') $\quad D + HCl(v=1) \xrightarrow{\text{a}} DCl + H$

$\xrightarrow{\text{b}} D + HCl(v=0)$

$$k_{3'a} = 10^{11,7} \text{ cm}^3 / \text{mol} \cdot \text{s}$$
$$\qquad T = 298 \text{K}$$
$$k_{3'b} = 10^{12,6} \text{ cm}^3 / \text{mol} \cdot \text{s}$$

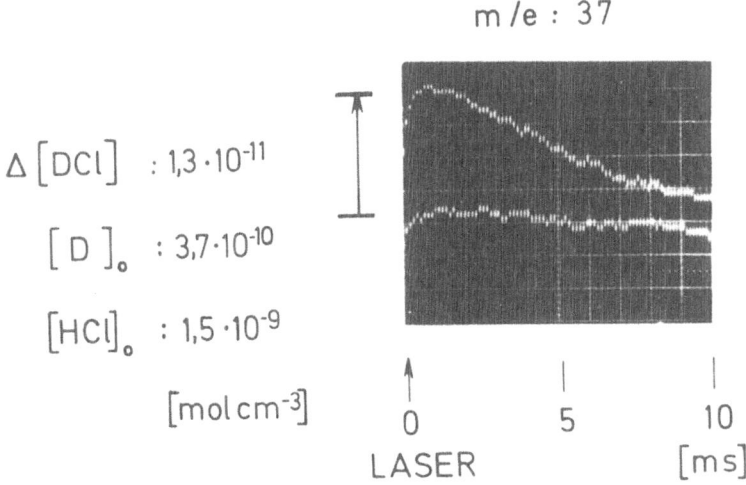

m/e : 37

$\Delta [DCl]$: $1,3 \cdot 10^{-11}$

$[D]_o$: $3,7 \cdot 10^{-10}$

$[HCl]_o$: $1,5 \cdot 10^{-9}$

$[\text{mol cm}^{-3}]$

0   5   10

LASER   [ms]

Fig. 8 Mass spectrometric detection of DCl formed in reaction (3'a)

In contrast to the $Cl + HCl(v)$ system the exchange reaction does not give the major contribution to the very fast vibrational deactivation [5,11,12]. A theoretical explanation of the mechanism of this observed effective nonreactive vibrational deactivation is still not available. Classical trajectory calculations predict a predominance of the chlorine exchange reaction [9, 13]. The reason for this appears to be the low potential energy barrier used in this calculations for the chlorine exchange path.

Recent experiments in discharge flow systems [14] and molecular beams [15, 16] as well as ab initio calculation [17, 18] are consistent with a higher barrier for chlorine exchange. It would also be very interesting to compare classical and quantum calculations for this system on sufficiently accurate potential energy surfaces, since quasi classical trajectories apparently greatly underestimate the importance of nonreactive inelastic collisions.

The interaction of $O(^3P)$ atoms with HCl

(4) $O(^3P) + HCl \longrightarrow OH + Cl \quad \triangle H^o_o = + 4,2\,kJ/mol$

$\quad k_4 = 10^{12,7} \exp(-3230/T)\ cm^3/mol \cdot s$

is also a nearly thermoneutral reaction with a potential energy barrier below $HCl(v=1)$.
As shown by the consumption of HCl in Fig. 9 this reaction is significantly accelerated by a single quantum excitation.

(4') $O(^3P) + HCl(v=1) \xrightarrow{\ a\ } OH + Cl$

$\qquad\qquad\qquad\qquad\quad \xrightarrow{\ b\ } O(^3P) + HCl(v=0)$

$\quad k_{4'a} = 10^{10,8}\ cm^3/mol \cdot s$

$\qquad\qquad\qquad\qquad\qquad\qquad T = 298K$

$\quad k_{4'b} = 10^{11,8}\ cm^3/mol \cdot s$

The rate enhancement factor of 10^2 is somewhat smaller than for reaction (2') and (3'). Similar to (3') relaxation predominates over reaction. A theoretical model to explain the effective energy transfer in collisions involving P-state atoms was given by NIKITIN [19]. As shown in Fig. 10 several potential surfaces exist for the interaction of $O(^3P)$ atoms with HCl. Each electronic state shows a repulsive interaction with a different shape. At certain distances these curves "cross" and a nondiabatic coupling between the different vibronic states is possible. However, detailed evaluations of these potentials are still not available.

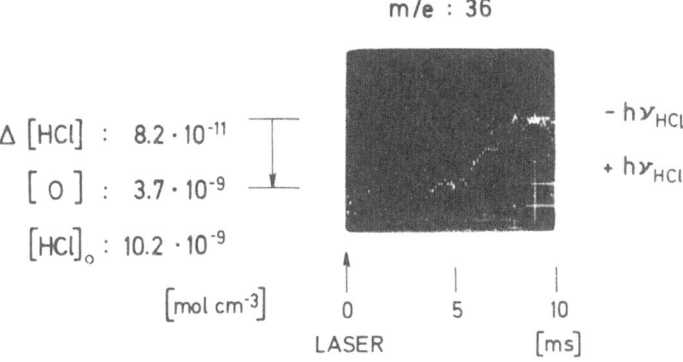

Fig. 9   Mass spectrometric measurement of the
          consumption of HCl in reaction (4' a)

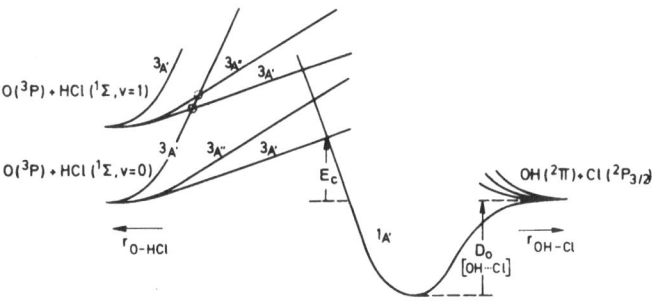

Fig. 10   Vibrational deactivation of $HCl(v=1)$ in
          electronically nonadiabatic collicions with
          $O(^3P)$ atoms according to NIKITIN [19]

4. Reactions of Vibrationally Excited Polyatomic Molecules

A spectacular method for excitation of polyatomic molecules into
very high vibrational levels has been discovered by observing the
selective dissociation of molecules after absorption of photons from
high power lasers which have a quantum energy more than an order
of magnitude smaller than the molecular dissociation energy [20].

A drawback of this procedure is the low amount of conversion compared with the large number of photons applied. Also at these high
energy levels near the dissociation threshold intramolecular energy
migration becomes very efficient [21 - 23]. From this point of view
chemical reactions involving polyatomic molecules in lower vibrational levels are especially interesting. In order to obtain information about the chemical reaction of a polyatomic molecules which
is vibrationally excited in a specific chemical bond one must try

decouple the vibration to vibration energy exchange from the removal of the excited molecule by interaction with the added reactant. By a low partial pressure of the excited molecules the time
for collisional energy transfer can be made long compared to the
time of reactive collisions. However, very rare information is
available on the intramolecular energy flow and reactions of polyatomic molecules in low vibrational levels. We have chosen substituted methanes for such investigations. Fig. 11 shows the level
diagram for methylfluoride up to 3000 cm^{-1}. At 9,55 μm a

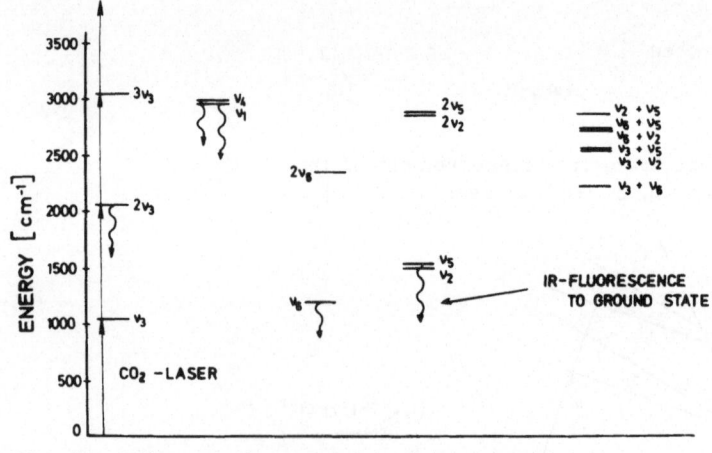

Fig. 11   Vibrational energy level diagram for
          CH_3F

coincidence between the P(20) line of the CO_2 laser and the C-F
stretching vibration (v_3) exist. By observing the infrared fluorescence from different vibrational levels the multiphoton excitation
of CH_3F with an unfocussed laser beam at moderate power levels
($\leq$ 100 kW/cm^2) could directly be monitored. Fig. 12 shows the
population of the C-H stretching mode (v_1, v_4) with an CH_3F partial pressure of 10 m Torr. Excitation of these levels is achieved
in much shorter times as calculated from previously measured rate
for the collisional vibration - to - vibration energy transfer [24].

Another way to a specific population of lower vibrational levels in
polyatomic molecules is the application of tunable infrared lasers.
As schematically shown for $CHCl_3$ in Fig. 13 two endothermic
reactions of substituted methanes with Br atoms require an overtone excitation of the C-H stretch vibration. This can be achieved by using a flashlamp dye laser as pump source for a temperature controlled $LiNbO_3$ crystal parametric oscillator with a
conversion efficiency of 15%. In connection with a multireflecting
mirror system (50 cm long) of a White design full absorption of
the infrared quanta is possible using $CHCl_3$ partial pressures

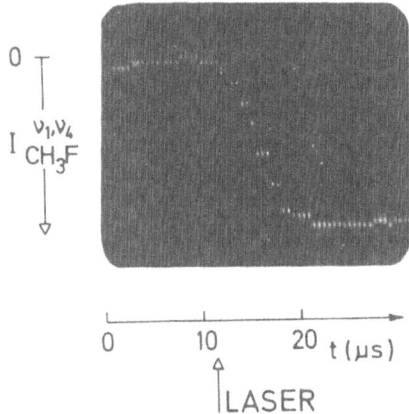

<u>Fig. 12</u>   Infrared fluorescence of CH_3F (v_1, v_4) at
3,4 µm after multiphoton excitation at
9,55 µm . P_{CH_3F} = 10 m torr

down to 1 Torr. Due to the dense line accumulation in the Q-branch
of the $CHCl_3(2 v_1)$ overtone band a laser line width of few tens of
a wavenumber is sufficient. A sensitive detection of the reaction
product HBr can be achieved using a molecular HBr laser and
the same light path as the excitation source for the absorption measurement. Replacing the HBr laser by a tunable laser diode will
give a sensitive double resonance arrangement for the study of a
large number of reactions of vibrationally excited polyatomic molecules.

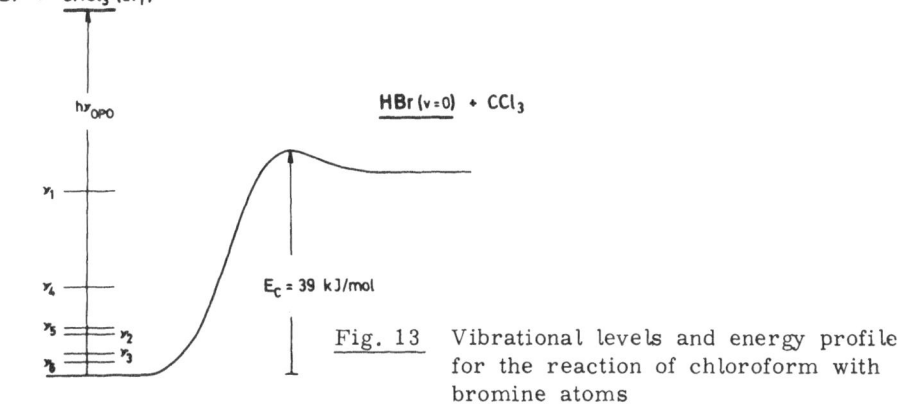

<u>Fig. 13</u>   Vibrational levels and energy profile
for the reaction of chloroform with
bromine atoms

206

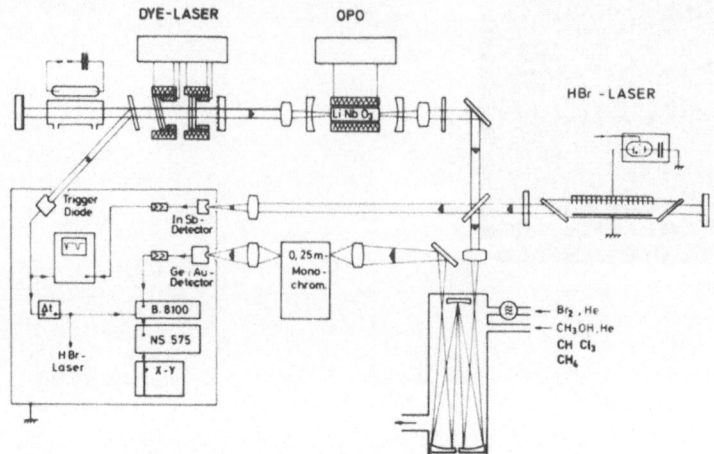

Fig. 14 Double resonance apparatus for the
investigations of reactions of vibrationally
excited polyatomic molecules with bromine
atoms

Acknowledgment

The authors wish to thank Prof. H.Gg. Wagner for his continuous
interest in this work and the Deutsche Forschungsgemeinschaft for
their financial support.

References

1. J. Wolfrum, Ber. Bunsenges. physik. Chem. 81, 114 (1977)
2. K. J. Schmatjko and J. Wolfrum, Ber. Bunsenges. physik.
   Chem. 79, 696 (1975)
3. C. L. Lin and F. Kaufman J. Chem. Phys. 55, 3760 (1971)
4. D. M. Hunten and M. McElroy, Rev. Geophys. 4, 303 (1966)
5. D. Arnoldi and J. Wolfrum, Ber. Bunsenges. physik. Chem.
   80, 892 (1976)
6. F. S. Klein, A. Persky and R. E. Weston, J. Chem. Phys.
   41, 1799 (1964)
7. P. N. Noble and G. C. Pimentel, J. Chem. Phys. 49, 3165 (1968)
8. D. G. Truhlar, P. C. Olson and C. A. Parr, J. Chem. Phys.
   57, 4479 (1972)
9. I. W. M. Smith, J. Chem. Soc. Faraday II, 71, 1970 (1975)
10. P. Botschwina and W. Meyer, Chem. Phys. Letters 44, 449 (1976)
11. R. G. Mcdonald, C. B. Moore, I. W. M. Smith and F. J. Wodar-
    czyk, J. Chem. Phys. 62, 2934 (1975)
12. J. F. Bott and R. F. Heidner III, J. Chem. Phys. 64, 1544 (1976)
13. D. L. Thompson, H. H. Suzukawa and L. M. Raff, J. Chem.
    Phys. 62, 4727 (1975); 64, 2269 (1976)

14. H. Endo and G. P. Glass, Chem. Phys. Letters 44, 180 (1976)
15. J. P. McDonald and D. R. Herschbach, J. Chem. Phys. 62, 4740 (1975)
16. J. P. Toennies (personal communication)
17. P. Botschwina and W. Meyer, Chem. Phys. 20, 43 (1977)
18. T. H. Dunning, J. Chem. Phys. (in press)
19. E. E. Nikitin and W. N. Kondratjev, Doklady Akad. Nauk SSSR. 212, 149 (1973)
    E.E. Nikitin in Physical Chemistry - An Advanced Treatise W. Jost (ed.), Vol VI A, chapter 4, Academie Press, New York (1975)
20. V. S. Letokhov, Physics Today p. 23, May (1977)
21. M. J. Coggliola, P. A. Schultz, Y. T. Lee and Y. R. Shen, Phys. Rev. Letters 38, 17 (1977)
22. R. A. Marcus, Ber. Bunsenges. physik. Chem. 81, 190 (1977)
23. W. L. Hase and C. S. Sloane, Ber. Bunsenges. physik. Chem. 81, 207 (1977)
24. Z. Karny, A. M. Ronn, E. Weitz and G. W. Flynn, Chem. Phys. Letters 17, 347 (1972)

Carbon Isotope Separation by Multiphoton Dissociation of CF$_3$I

S. Bittenson and P.L. Houston

Department of Chemistry, Cornell University
Ithaca, NY 14853, USA

A selective multiphoton dissociation process has been used to enrich carbon-13 in CF$_3$I. A separation factor of nearly 600 has been achieved for irradiation of 0.10 torr of CF$_3$I at -80°C with the R(14) line of the 9.6 μ CO$_2$ laser transition. An investigation of the dependence of the enrichment factor on pressure indicates that collisions during the dissociation are effective in destroying the selectivity. The multiphoton dissociation is quite efficient. At laser intensities of 10 MW/cm^2, one in every 11 absorbed photons contributes its energy to the breaking of the C-I bond.

CF$_3$I dissociation was achieved with a grating tuned CO$_2$ TEA laser (Tachisto Corporation model 215 laser head) producing a maximum of 1 joule single line output in 60 ns. fwhm. A 30 cm focal length sodium chloride lens was used to focus the radiation through polished NaCl windows into cylindrical Pyrex sample cells. The dimensions of the cells were adjusted to meet requirements of individual experiments. Cell lengths ranged from 5 cm to 30 cm when focusing was used, and from 5 cm to 114 cm when the laser was used unfocused. Species identification, concentrations, and isotope ratios were determined with a Perkin Elmer model 521 grating infrared spectrometer and a Consolidated Engineering Corporation type 21-103A mass spectrometer.

Laser power was measured with a Scientech model 360001 laser power meter. The pulse intensity was taken to be one-half of the measured energy per pulse in 60 nanoseconds over the mean irradiated area of a sample. Beam areas were recorded on thermal sensing paper stock and were not corrected for laser mode structure or external diffraction effects. An intracavity aperture near the output mirror was used to restrict lasing to low order transverse modes, and an external aperture was generally used to reduce the beam area to 0.5 cm^2.

The CF$_3$ and I radicals formed by multiphoton dissociation of CF$_3$I recombine to yield CF$_3$I, C$_2$F$_6$ and I$_2$. These are the only species observed in either infrared or mass spectra of samples irradiated at intensities below 25 MW/cm^2. In particular, CF$_2$I$_2$ and other products which arise from breaking a C-F bond are not observed.

We define β_p to be the ^{12}C/^{13}C isotope ratio observed in the C$_2$F$_6$ product divided by the ratio we would expect for a nonselective process. Figure 1 shows the variation of β_p with the pressure of CF$_3$I. At high pressures $\beta \approx 1$ and no selectivity is observed. However, β_p increases

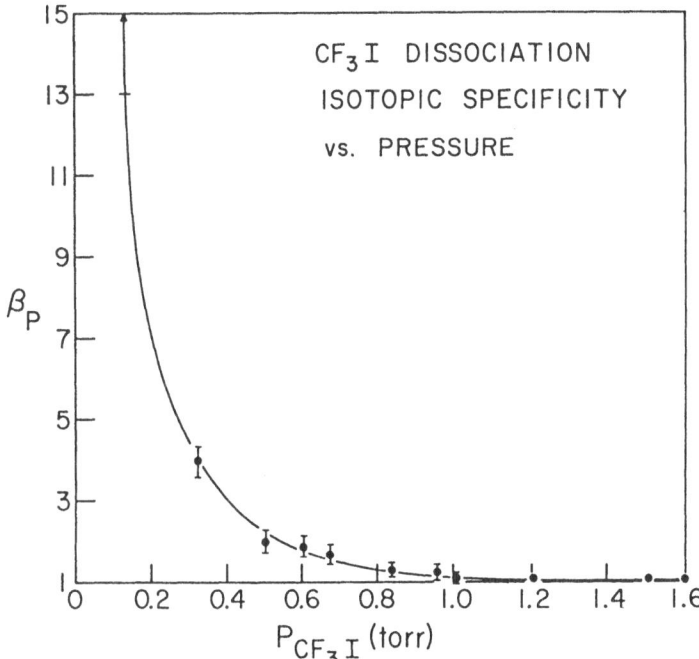

CF$_3$I DISSOCIATION
ISOTOPIC SPECIFICITY
vs. PRESSURE

β_p

P_{CF_3I} (torr)

Fig. 1 Selectivity of product formation. The laser intensity was 5.5 MW/cm^2.

dramatically as the pressure of CF$_3$I is reduced below 1.0 torr. Consequently, the CO$_2$ laser (R(14), 9.3 μ) selectively dissociates ^{12}CF$_3$I. One explanation for the dramatic change in β_p with pressure is that, at pressures above 1.0 torr, collisions during the laser pulse exchange energy between the selected ^{12}CF$_3$I and nonselected ^{13}CF$_3$I.

As the products of the multiphoton dissociation of CF$_3$I become enriched in ^{12}C, the remaining reactants become enriched in ^{13}C. We have achieved a 590-fold enrichment of ^{13}CF$_3$I by irradiating naturally occurring CF$_3$I (^{13}C/^{12}C 1/99) at 0.1 torr with 2000 laser pulses on the R(14) line of the 9.6 μ band. The peak intensity was 25 MW/cm^2 and the cell temperature was -80°C. Mass spectral analysis showed that 86% of the residual reactant was ^{13}CF$_3$I.

By measuring both the amount of energy absorbed by the sample and the number of product molecules, it is possible to determine the quantum efficiency of the multiphoton dissociation. We have examined this efficiency as a function of incident intensity over the range 1-10 MW/cm^2.

A substantial uncertainty in the efficiency arises from the kinetics. If CF$_3$ and I radicals recombine to an appreciable extent, or if CF$_3$ reacts with

I_2 accumulated from previous pulses, then the net number of products created or reactants consumed will be substantially less than the number of dissociations. This will result in a lower apparent efficiency for dissociation than that which would be obtained in the absence of these reactions. The fractional dissociation per shot and the efficiency reported below have not been corrected for these effects and should be viewed as lower bounds.

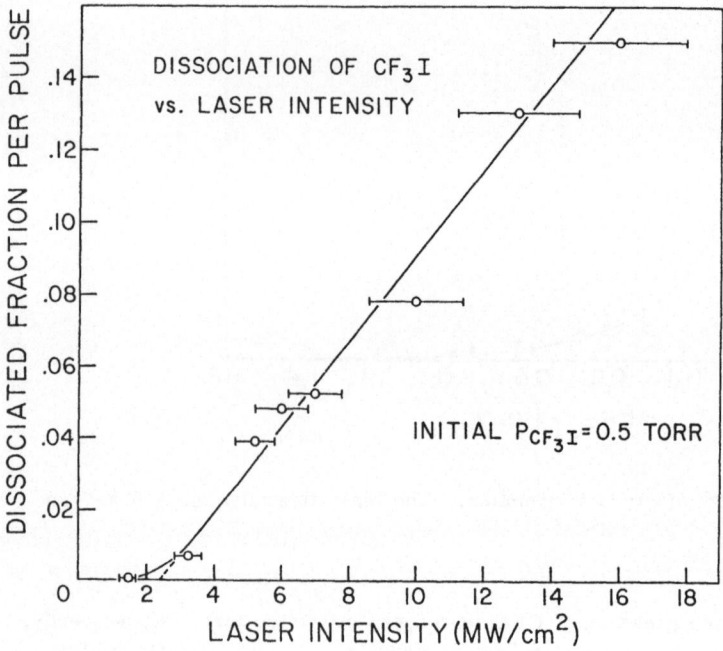

Fig. 2 Fractional dissociation of CF_3I in the beam per pulse as a function of laser intensity.

Figure 2 displays the fraction of molecules dissociated in the beam per pulse as a function of incident laser intensity. The intensity axis may be converted to energy fluence in J/cm^2 by multiplication by 0.12. The fractional dissociation exhibits a threshold below 2 MW/cm^2 and then increases with intensity. At an intensity of 10 MW/cm^2, roughly 8.4% of the molecules in the beam are dissociated per pulse, while at 16 MW/cm^2 the fraction increases to 15%.

The fractional dissociation in the beam per pulse may be converted to an efficiency if the amount of energy absorbed by the sample is measured. For 0.5 torr of CF_3I absorption could conveniently be measured in the intensity range below 11 MW/cm^2. At 10 MW/cm^2 16.6 photons were absorbed by the sample per molecule in the beam. Since the CF$_3$-I bond strength is 55 kcal/-mole and roughly equal to the energy of 18 laser photons, we define an efficiency of 1.0 to correspond to one dissociation for every 18 photons absorbed. Division of 8.4% by 16.6 and multiplication by 18 yields 9.1%, the efficiency of dissociation at 10 MW/cm^2. Consequently, at this intensity one in every 11 absorbed photons contributes its energy to the dissociation process. The actual efficiency may be even higher if CF$_3$I is reformed after dissociation.

Acknowledgement: We greatly acknowledge support for this research through a grant from the Standard Oil Company of Ohio.

Infrared Laser Dissociation of Tetramethyldioxetane

Y. Haas and G. Yahav

Department of Physical Chemistry, The Hebrew University of Jerusalem
Jerusalem, Israel

Dioxetanes [1] are four membered ring compounds that are known to decompose thermally according to (1)

$$
\begin{array}{c} \text{O-O} \\ |\ | \\ \text{-C-C-} \\ |\ | \end{array} \xrightarrow{\text{heat}} \begin{array}{c} \text{O} \\ || \\ \text{C} \\ /\backslash \end{array} + \begin{array}{c} \text{O} \\ || \\ \text{C} \\ /\backslash \end{array} + \text{visible light} \tag{1}
$$

The products are only carbonyl compounds, and light emission (at about 420 nm) is observed in both gas phase and liquid solution. The nature of the electronic excited states involved is not quite clear, as the chemiluminescence spectrum does not agree with either the singlet or the triplet of emission of the products. In solution, the species induces reactions that are usually brought upon by triplet sensitization. We undertook the study of high power infrared laser thermolysis of tetramethyldioxetane for 3 main reasons:
1) The time resolution of about 100 nsec allows the highest time resolution yet applied to the reactions.
2) The system promises to be a good probe for the study of the mechanism of multiphoton dissociation by infrared lasers, [2]. As the reaction is accompanied by visible light emission, one has a very sensitive way to monitor the real time evolution of the reaction.
3) The system was mentioned as a possible visible laser candidate. The laser excitation method should afford a high concentration of excited states in a short time, thus allowing evaluation of rate constants, transient absorption characteristics, gain or loss, etc.

The laser used for this study is commercial (El-O_p Industries, Rehovot, Israel) TEA laser, operated on the R(24) or R(26) of the 00^01-10^00 transition. Pulse energy was about 0.5 Joule with two thirds of the energy contained in a spike 80 nsec FWHM, the rest is a "tail" lasting for about 1.5 μsec. The time resolved emission at 420 nm for a 70 mtorr sample is shown in Fig.1. It is seen that initially the emission follows closely the laser spike, then there is a slow (few hundred nanoseconds) rise in emission intensity, and finally a slower decay. At higher pressures one observes a fast rise (simultaneous with the laser pulse) and then a decay. No delay between the laser spike and the chemiluminescence is observed in either case. The chemiluminescence spectrum is shown in Fig.2 along with the fluorescence and phosphorescence spectra of acetone, the only product. It is clear that the chemiluminescence spectrum does not coincide with either one.

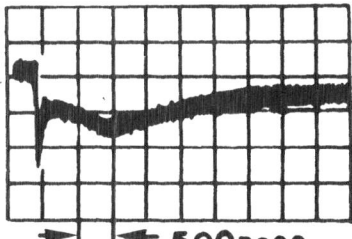

Fig.1 Oscilloscope trace of the chemiluminescence of tetramethyldioxetane at 0.07 torr

It shows good agreement, however, with chemiluminescence spectra obtained by conventional heating of tetramethyldioxetane in both gas phase and liquid solution.

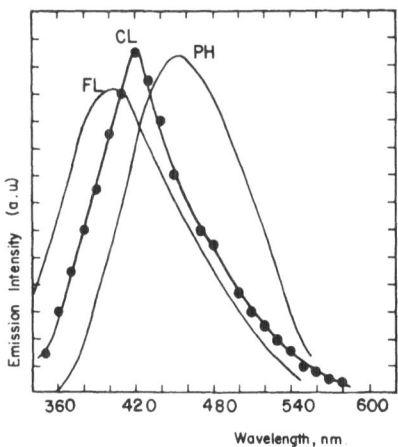

Fig.2 The chemiluminescence spectra of tetramethyl-dioxetane at 0.35 torr. Fluorescence and phosphorescence spectra are shown for comparison

Our results may be explained tentatively by assuming the following pathway for dioxetane laser thermolysis, (2)-(6):

$$D + nh\nu(IR) \rightarrow D^{\ddagger} \tag{2}$$

$$D + n'h\nu(IR) \rightarrow D^{+} \tag{3}$$

$$D^{+} + D^{+} \rightarrow D^{\ddagger} + D \tag{4}$$

$$D^{\ddagger} \rightarrow \text{product} + h\nu(\text{vis}) \tag{5}$$

$$D^{+}(D^{\ddagger}) + M \rightarrow D + M \tag{6}$$

Laser excitation creates, depending on the number of photons absorbed, either superexcited molecules (designated $D^{\ddagger}$) that decompose instantaneously, or vibrationally excited ones (D^{+}) not containing quite enough energy to dissociate. These mildly excited molecules may acquire the energy needed to decompose by collision (4), or lose energy by collisional deactivation (6). At low pressures one observes the initial creation of $D^{\ddagger}$ by the chemiluminescent spike following closely the laser spike, and then process (4) by the subsequent slow rise in emission intensity. At higher pressures these two mechanisms blend together due to the higher collision frequency. Total deactivation of the vibrationally excited population accounts for the final slow decay, which is faster the higher the pressure.

As to the nature of the emitting species - we believe our results rule out the triplet state of acetone. This follows from comparison of the relevant spectra, and also from kinetic considerations - the triplet is known to be long lived (170 microseconds) and the emission we observe is at least 2 orders of magnitude shorter.

Since singlet emission may also be ruled out by spectroscopic data, we suggest the possibility of excimer emission, an option mentioned by Kopecky in the first paper on the subject [3]. The red shift compared to the singlet is normal with excimer emission, as well as the short lifetime. Unfortunately no excimer emission from acetone has been yet reported, so that direct comparison cannot be made for now. As to the triplet type behaviour in inducing chemical reactions - it could be that the excimer behaves as a triplet in that respect, or that it partially decomposes to yield a triplet state.

References

1. T. Wilson, in D.R. Herschbach, Ed., International Review of Science, Series 2, Vol. 9, "Chemical Kinetics", Butterworth 1976, p. 265.

2. R.V. Ambartzumian, N.V. Chekalin, Y.A. Gorokhov, V.S. Letokhov, G.N. Makarov and E.A. Ryabov, Laser Spectroscopy, Proceedings of the Second International Conference, Megeve, Springer Verlag 1975, p. 121.

3. K.R. Kopecky and C. Mumford, Can. J. Chem. **47** (1969) 709.

Observation of Tunable-Laser-Induced Grating Dip

F. Keilmann

Max-Planck-Institut für Festkörperforschung
7000 Stuttgart, FRG

Abstract
Fine structure in form of a central dip was found in the saturation spectrum obtained in a two-wave experiment. This effect con-
firms theoretical predictions. It originates from spatial hole-
burning.

Introduction
High power lasers are essential for performing saturation spectroscopic investigations in condensed media. The reason for this
is that due to strong nonradiative interactions the relaxation
of excited states occurs rapidly, typically on the time scale
of picoseconds.
    We have recently studied intervalence band electronic tran-
sitions in room-temperature germanium using CO_2 laser radiation
at 40 MW/cm^2 [1]. The result was a laser-induced reduction of
the absorption extending over a spectral width of about 100 cm^{-1}
(FWHM), much smaller than the width of the absorption continuum.
We concluded that this transition can be regarded as inhomoge-
neously broadened with a homogeneous width of 100 cm^{-1}. From
this a corresponding dipole lifetime T_2 = 0.1 ps results which
finds quantitative explanation by considering scattering of the
excited carriers by an optical phonon of the lattice.
    However, for fully relaxing the excited carrier back to the
ground state, a cascade of further scattering processes has to
follow. Thus a relatively long population equilibration time results which we call T_1 in analogy to the terminology used for
two-level systems. With $T_1 \gg T_2$ we have a system capable of displaying the tunable-laser-induced grating dip on the bottom of
the hole-burning lineshape in a two-wave saturation experiment.

Grating dip
When two waves nonlinearly interact in a two-level saturable absorber the interaction cannot be fully described by considering
level populations only. Additional contributions come from a
coherent superposition of ground and excited state. They lead to
a modification of the lineshape observed.
    For example, consider an inhomogeneously broadened transition
partially bleached by a strong radiation ("saturator"). The hole
burnt is a Lorentzian displaying the homogeneous width times a
power broadening factor. This lineshape is observable, e.g., by
looking at the spontaneous emission from the excited state.
    If, however, a second, weak radiation ("probe") is used to

simultaneously measure the hole-burning effect, a markedly different lineshape results from the above-mentioned coherent contributions. This has been calculated by BAKLANOV and CHEBOTAEV [2], SARGENT III [3] and Sargent III and TOSCHEK [4]. To our knowledge it has previously not been observed.

Qualitatively, the lineshape modifications can be understood using the classical picture of the "laser grating" [3]. As shown in Fig. 1 the two intersecting waves give rise by interference, to weak-contrast fringes which move at a speed proportional to the frequency difference of both waves. This moving intensity grating slaves with it a population difference grating since the intensity is in the saturating regime. In turn a refractive index grating is coupled to the population grating, also with weak contrast if the probe is weak.

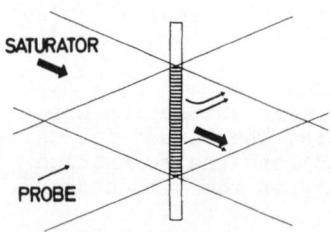

Fig. 1 Schematic view of two-wave saturation experiment. By interference of the saturating and probing waves a spatial and temporal modulation is generated in the level populations of the medium ("grating"). This in turn scatters light from the saturating beam into the forward direction of the probe beam and vice versa, as symbolized by the bent arrows

This probe induced grating now scatters some saturator intensity into the probe direction and vice versa (Bragg scattering). Since the grating moves a Doppler shift occurs in the scattering. This automatically matches the scattered (bent arrows in Fig. 1) to the direct (straight arrows) part so that they become undistinguishable. Observation of the probe's "transmission" under these circumstances, for zero difference frequency, yields a relative absorption coefficient [4]

$$\frac{\alpha}{\alpha_0} = 1 + \frac{1-\sqrt{1+I/I_s}}{\sqrt{1+I/I_s}} - \frac{I/I_s}{2(1+I/I_s)^{3/2}}$$

where α_0 is the small signal absorption coefficient, I the saturator intensity, and I_s the saturation intensity of the medium. The second term is the hole-burning contribution which is approximately Lorentzian with a width (FWHM) of $(1+\sqrt{1+I/I_s})$ / (πT_2). The third term comes from the grating. Note that at zero difference frequency both terms are of the same order of magnitude.

The grating dip now occurs in cases when the third term vanishes on tuning to relatively small difference frequencies. This is fulfilled in a medium with $T_1 \gg T_2$. Consider a difference frequency $1/2\pi T_1$ well inside the broad hole-burning Lorentzian. At this point the medium's population difference is modulated at its relaxation rate, and the forced oscillation is therefore damped. At higher difference frequencies the population grating fully washes out and the third term vanishes. The grating dip has approximately Lorentzian shape with a width of

$$\sqrt{1+I/I_s} \ / (\pi T_1) \text{(FWHM)} \quad [4].$$

Experiment in p-Germanium
The double-dip saturated lineshape was calculated for the intervalence band transition in germanium by SARGENT III [5]. Our experiment [6] involved a strong single ns pulse at 942 cm^{-1} to induce saturation and a weak, relatively long pulse for probing. The frequency of the latter was step-tunable between 931 and 1086 cm^{-1}. The transmission of the probe light was observed to be enhanced during the passage of the saturating pulse. From the enhancement factor an effective absorption coefficient was evaluated and plotted vs. the probe frequency (Fig. 2).

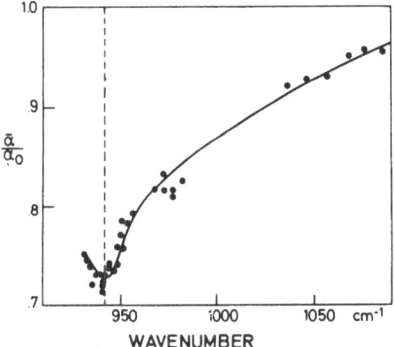

Fig. 2 Saturated probe absorption spectrum of p-Ge in presence of strong radiation at 942 cm^{-1}. The grating dip, approx. 20 cm^{-1} in width, is seen on the broad hole-burning background.

The resulting spectrum clearly shows the grating dip at the center of the broad hole-burning dip. We can read the two linewidths, approximately, to be 20 and 210 cm^{-1}. An estimation of the (small) power broadening factors can be made from the depth of the saturation at 942 cm^{-1} using the formula above. Thus we can take two lifetimes $T_1 = .6ps$ and $T_2 = .1ps$ from the spectrum in Fig. 2.

Applications

The grating dip can be used for a nearly direct determination of the population difference relaxation time. This is especially important in the study of rapid relaxation in condensed media. No more direct method exists for lifetimes in the sub-picosecond region. On the other side, grating dips should be observable in gaseous media. Cases might be found where the grating dip is much narrower than the Lamb dip.

References

1 F. Keilmann, IEEE J. Quant. Electr. QE 12, 592 (1976)
2 E.V. Baklanov and V.P. Chebotaev, Sov. Phys. JETP, 34 490 (1972)
3 M. Sargent III, Appl. Phys. 9, 127 (1976)
4 M. Sargent III and P.E. Toschek, Appl. Phys. 11, 107 (1976)
5 M. Sargent III, Opt. Comm. 20, 298 (1977)
6 F. Keilmann, Appl. Phys. 14, 29 (1977)

Study of the Visible Fluorescence of Gaseous UF$_6$

A. Andreoni, R. Cubeddu, S. De Silvestri, and F. Zaraga

Centro di Studio per l'Elettronica Quantistica del C.N.R.
Istituto di Fisica del Politecnico
Milano, Italy

The visible fluorescence from the forbidden A-X absorption band
(340-410 nm) of gaseous UF$_6$ has been observed previously by differefent authors [1, 2, 3]. The lifetime at p=0 turns out to be
considerably shorter than that estimated from the integrated
absorption spectrum, indicating a non-radiative process as the
agent mainly responsible for unimolecular decay. In addition, a
quenching rate at room temperature corresponding to a high number of collisions has been measured. We report here further investigations on gaseous UF$_6$ excited at 374 and 270 nm. As regards
the excitation at 374 nm, we have studied the dependence of the
fluorescence decay time on pressure at different temperatures.

The experimental apparatus was similar to that previously
described [2]. The fluorescence curve, after a fast initial transient, can be approximated to by an exponential decay law with
a single time constant τ.

The reciprocal lifetime $1/\tau$ vs pressure for different temperatures, ranging from 273 to 313 $^\circ$K, are shown in Fig.1. The experimental points are well fitted by straight lines, according to
the Stern-Volmer equation:

$$\frac{1}{\tau} = \frac{1}{\tau_0} + K_q\, p \tag{1}$$

Values of the quenching rate K_q (slope of the lines in Fig.1)
and of unimolecular decay time τ_0 (p=0 extrapolation of the decay
times) are reported in Table 1 for different temperatures. The
standard deviations on K_q are also reported. They account for
the statistical fluctuations due to noise in each fluorescence
waveform and do not take into account either errors in the measurement of pressure and temperature, nor uncertainties in the
background subtraction. The quenching rate is shown to increase
very fast with the temperature. This implies that the number of
collisions effective for quenching changes with the temperature.
Such a behaviour can be explained by assuming the probability
of quenching to be zero for collisions between molecules whose
kinetic energy of mutual approach is less than a critical energy
E_{cr} and unity for kinetic energy greater than E_{cr}. The quenching
rate K_q for an ideal gas in the hard sphere approximation, neglecting the internal degrees of freedom of the molecules, is

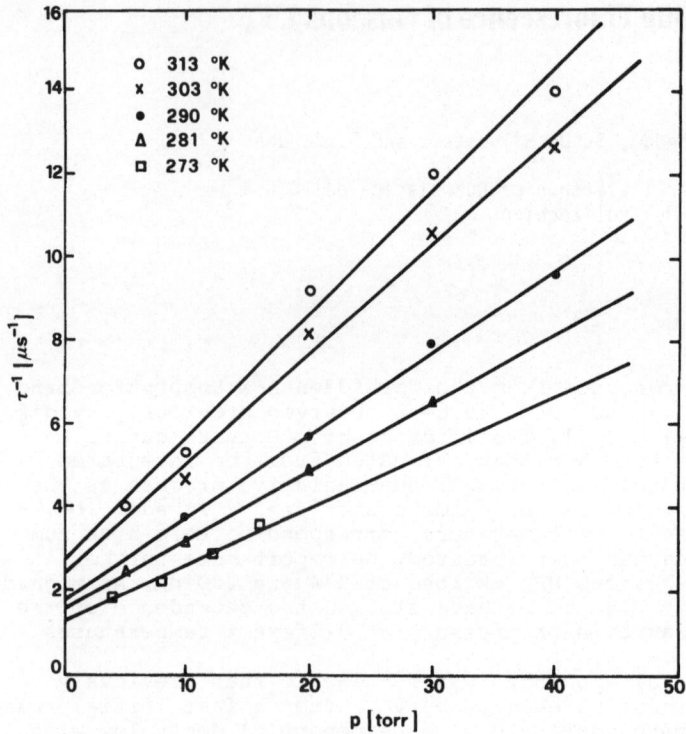

Fig.1 Stern-Volmer plot of the reciprocal lifetime vs pressure for different temperatures

Table 1

T [°K]	K_q [torr^{-1} μs^{-1}]	τ_0 [ns]
273	0.137 ± 0.007	775
281	0.169 ± 0.007	639
290	0.198 ± 0.012	552
303	0.263 ± 0.006	401
313	0.295 ± 0.008	371

then given by:

$$K_q = A \exp (-E_{cr}/ kT) \tag{2}$$

where

$$A = \frac{\pi D^2 \langle v \rangle}{kT} \tag{3}$$

πD^2 is the collisional cross section, $\langle v \rangle$ is the average molecular velocity, k is the Boltzmann's constant and T is the absolute temperature. The steric factor does not appear in (3) since the critical energy was assumed to be independent of the mutual orientation between the colliding molecules. Figure 2 shows a semi-logarithmic plot of the experimental values of K_q vs the reciprocal temperature.

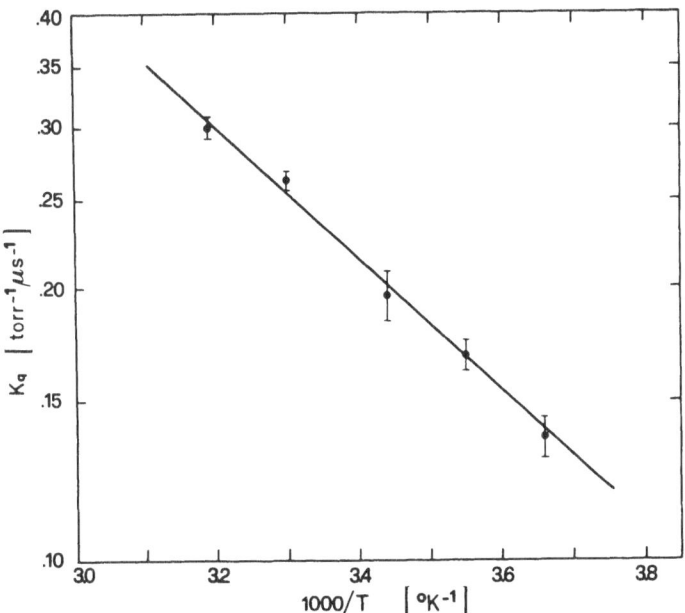

Fig.2 Arrhenius plot of the quenching rate vs the reciprocal temperature

As expected from (2), the experimental points are well fitted by a straight line in the temperature range considered. The interpolating line drawn in Fig.2 corresponds to a critical energy $E_{cr}=1150$ cm^{-1} and to an Arrhenius pre-exponential factor A=60 torr^{-1} μs^{-1}. This value of A, once substituted into (3), gives D=20 Å at room temperature, in reasonable agreement with the gas-kinetic collisional diameter D as calculated from (31) of Ref.[4] (8.4 Å). The experimental data do not allow us to discriminate amongst the possible mechanisms of quenching. It must be noted, however, that the energy of the excitation photons in our experiment (26740 cm^{-1}), plus the critical energy, give ~80 Kcal/ /mole, which is approximately the value of the dissociation energy reported in the literature [5] for the reaction:

$$UF_6 \rightarrow UF_5 + F$$

This could be an indication that the quenching process is colli
sion-induced dissociation following the absorption of a single
photon. The unimolecular decay time τ_0 also decreases with increasing temperatures (see Table 1). Due to the uncertainty of
the extrapolation at p=0, it is difficult to establish a defi
nite law for this dependence. Two physical mechanisms may be
responsible for this behaviour: predissociation and internal con
version.

The strong allowed $\tilde{B}$-$\tilde{X}$ absorption band (250-300 nm) has been
studied by means of the 270 nm radiation from the second harmo
nic of a Disodiumfluorescein dye laser. The observation band was
only limited by the spectral response of the photomultiplier
(S-11 surface). This absorption band has been assigned by LEWIS
et al. [6] to the transitions $\gamma_{8u} \rightarrow \gamma_{8u}'$, $\gamma_{6u} \rightarrow \gamma_{8u}$, $\gamma_{8u} \rightarrow \gamma_{7u}'$, and
the structure in the band has been attributed to vibrational pro
gressions [6]. Despite this indication that the electronic state
should be bounded, no fluorescence at all has been detected with
a sensitivity of 0.02 fluorescence photons per ns. Since the ab
sorption cross section at 270 nm is about fifty times higher
than the one at 374 nm [7], this shows the existence of a very
efficient non-radiative process.

This research was partially supported by the Italian C.N.R.
and performed under a contract between C.N.E.N. and Politecnico.

REFERENCES

1 A.Andreoni and H.Bücher: Chem.Phys.Lett. <u>40</u>, 237 (1976)

2 P.Benetti, R.Cubeddu, C.A.Sacchi, O.Svelto and F.Zaraga:
   Chem.Phys.Lett. <u>40</u>, 240 (1976)

3 O.De Witte, R.Dumanchin, M.Michon and J.Chatelet: Chem.Phys.
   Lett. <u>48</u>, 505 (1977)

4 R.De Witt: In <u>Uranium Hexafluoride: A Survey of the Physico-
   Chemical Properties</u> (Goodyear Atomic Corporation Report GAT-
   -280, 1960) p.49

5 N.P.Galkin and Yu.N.Tumanov: Atomnaya Energiya <u>30</u>, 372 (1971)

6 W.B.Lewis, L.B.Asprey, L.H.Jones, R.S.McDowell, S.W.Rabideau
 . and A.H.Zeltmann: J.Chem.Phys. <u>65</u>, 2707 (1 76)

7 R.McDiarmid: J.Chem.Phys. <u>65</u>, 168 (1976)

List of Participants

Graham S. Arnold
Room 2-045, MIT
77 Massachusetts Ave.
Cambridge, Ma 02139
USA

F. Aussenegg
Physikalisches Institut der
Universität Graz
Universitätsplatz 5
8010 Graz
Austria

Mario Bertolotti
Universitá di Roma
Facoltá di Ingegneria
Istituto di Fisica
P.le Scienze 5
Roma
Italy

Willy L. Bohn
Deutsche Forschungs- und Versuchs-
anstalt für Luft- und Raumfahrt e.V.
Pfaffenwaldring 38 - 40
7000 Stuttgart 80
Germany

D.J. Bradley
Imperial College of Science
and Technology
Optics Section
London SW 7 2BZ
England

Ch.A. Brau
University of California
Los Alamos Scientific Laboratory
P.O.Box 1663
Los Alamos, NM 87545
USA

Günter Brederlow
Projektgruppe für Laserforschung
der Max-Planck-Gesellschaft
8046 Garching
Germany

Hermann Bücher
Lambda Physik
Wagenstieg 8
34 Göttingen
Germany

Ronald L. Bullock
R 1/1178
TRW Systems
Redondo Beach, Calif. 90278
USA

George Burns
Department of Chemistry
University of Toronto
80 St. George St.
Toronto
Canada

L.F. Champagne
N.R.L.
Code 5540
Washington D.C. 20375
USA

Paul Christensen
Department of Electrical Engineering
University of California
Los Angeles, Calif. 90007
USA

George Collins
Dept. of Electrical Engineering
Colorado State University
Fort Collin, Colorado
USA

Rinaldo Cubeddu
Via Piave 21
Milano
Italy

Edward Danielewicz
Institut für Festkörperforschung der
Max-Planck-Gesellschaft
Büsnauer Str. 171
7000 Stuttgart 80
Germany

N. Donklias
Siemens AG
Balanstr. 73
8000 München 90
Germany

N. Djeu
US Naval Research Laboratory
Washington, D.C. 20390
USA

Jacques Ducuing
Laboratoire d'Optique du C.N.R.S.
Ecole Polytechnique
91128 Palaiseau Cedex
France

R. Evenkamp
Treufer Str. 2
85 Nürnberg
Germany

Viktor Evtuhov
Hughes Research Laboratory
3011 Malibu Canyon Road
Malibu, California
USA

R.W. Field
Massachusetts Institute of
Technology
Department of Chemistry
Cambridge, Mass. 02130
USA

Sieghart Fischer
Technische Universität
Department Physik T 30
James-Frank-Str.
8046 Garching
Germany

Bernard Fontaine
Institut de Mécanique des Fluides
de Marseille
1, rue Honnorat
13003 Marseille
France

K.D. Foster
DREV
17 Chemin du Plateau
Lac Beau Port
Quebec
Canada

Hubert Guillet
13, rue Lamartine
91220 Bretigny sur Orge
France

James H. Gorrell
EOARD
223/231 Old Marylebone Road
London NW1 5TH
England

Günter Haag
Institut für Theoretische Physik der
Universität
Pfaffenwaldring 57
7000 Stuttgart 80
Germany

Dieter Haaks
G.H. Wuppertal
Physikalische Chemie
Gaußstraße
Wuppertal 1
Germany

Yehuda Haas
Dept. of Physical Chemistry
The Hebrew University
Jerusalem
Israel

Richard Heidmann
36, allée circulaire
27200 Vernon
France

Hanspeter Helm
Institut für Atomphysik der Univer-
sität Innsbruck
Karl Schönherrstr. 3
6020 Innsbruck
Austria

Wolfgang Heudorfer
Institut für Theoretische Physik
der Universität
Pfaffenwaldring 57
7000 Stuttgart 80
Germany

Peter Hoffmann
Pfaffenwaldring 38
7000 Stuttgart 80
Germany

R. Hofland
Aerospace Corporation
Los Angeles, Calif. 90009
USA

Kristian Hohla
Projektgruppe für Laserforschung
der Max-Planck-Gesellschaft
8046 Garching
Germany

Paul Lyon Houston
Dept. of Chemistry
Cornell University
Ithaca, NY 14853
USA

Helmut Hügel
DFVLR - Institut für Technische
Physik
Pfaffenwaldring 38
7000 Stuttgart 80
Germany

F. Keilmann
Institut für Festkörperforschung
der Max-Planck-Gesellschaft
Büsnauer Str. 171
7000 Stuttgart 80
Germany

T.A. King
Physics Department
Schuster Laboratory
Manchester University
Manchester M13 9PL
England

Georg Kinshofer
Lilienstraße 85
8011 Vaterstetten
Germany

Alan F. Klein
Systems, Science, Software
P.O.Box 4803
Hayward, Calif. 94540
USA

Claude A. Klein
Raytheon Research
Waltham, Mass. 02154
USA

Ekkehard Klement
Siemens AG
ZFE FL OPT 11
Postfach 700076
8000 München 70
Germany

Edgar Klose
Akademie der Wissenschaften der DDR
Zentralinstitut für Optik und
Spektroskopie
Rudower Chaussee 6
1199 Berlin-Adlershof
DDR

Karl Ludwig Kompa
Projektgruppe für Laserforschung der
Max-Planck-Gesellschaft
8046 Garching
Germany

Jürgen Kuhl
Institut für Festkörperforschung
der Max-Planck-Gesellschaft
Büsnauer Str. 171
7000 Stuttgart 80
Germany

H.R. Lüthi
Institut für Angewandte Physik
der Universität
Sidlerstr. 5
3012 Bern
Schweiz

Joseph A. Mangano
Avco Everett Research Lab. Inc.
Everett, MA 02149
USA

Marowsky
Institut für biophysikalische Chemie
der Max-Planck-Gesellschaft
Postfach 968
3400 Göttingen-Nikolausberg
Germany

Richard Meinzer
United Technologies Research Center
Hartford, CT 06067
USA

Robert Meredith
12 Research Dr.
Ann Arbor, Michigan
USA

G.L. Oomen
Physics Department
Twente University of Technology
P.O.Box 217
Enschede
Netherlands

T.D. Padrick
Sandia Laboratories
Albuquerque, NM 87115
USA

Joel H. Parks
AVCO Everett Research Lab. Inc.
2385 Revere Beach Parkway
Everett, Mass. 02149
USA

P.J. Peters
Physics Department
Twente University of Technology
P.O.Box 217
Enschede
Netherlands

Howard T. Powell
Lawrence Livermore Laboratory
L-470
Livermore, Calif.
USA

Alfred T. Pritt, jr.
Science Center
Rockwell International
P.O.Box 1085
Thousand Oaks, Calif.
USA

Gabrielle Kovarna Raff
Tachisto Incorporated
13 Highland Circle
Needham, MA 02194
USA

Eggo Ratsch
Lauterstr. 38
1000 Berlin 41
Germany

Howard Rausch
Laser Focus
385 Elliot Street
Newton, MA 02164
USA

Len Reed
164, Commercial St.
Sunnyvale, Calif. 94086
USA

B.J. Reits
Physics Department
Twente University of Technology
P.O.Box 217
Enschede
Netherlands

C.K. Rhodes
Stanford Research Institute
Menlo Park, Calif. 94025
USA

G.L. Rogoff
Westinghouse R & D. Center
Beulah Road
Pittsburgh, PA 15235
USA

J.P. Ryan
Optics Group
Blackett Laboratory
Imperial College
London SW7 2BZ
England

Rainer Salomaa
Helsinki University of Technology
Dept. of Physics
SF-02150 Espoo 15
Finland

Fritz Peter Schäfer
Max-Planck-Institut für biophysika-
lische Chemie
Postfach 968
3400 Göttingen
Germany

Wolfram E. Schmid
Projektgruppe für Laserforschung der
Max-Planck-Gesellschaft
8046 Garching
Germany

Murray Sargent
Optical Sciences Center
University of Arizona
Tucson, Arizona 85721
USA

Wolfgang Seelig
Institut für Angewandte Physik
der Universität
Sidlerstr. 5
3000 Bern
Schweiz

Wolfgang Schall
DFVLR -Institut für Technische
Physik
Pfaffenwaldring 38
7000 Stuttgart 80
Germany

Roland Schmiedl
Fakultät für Physik der
Universität Bielefeld
48 Bielefeld
Germany

Bernd Schramm
Phys.-Chem. Institut
Im Neuenheimer Feld 253
6900 Heidelberg
Germany

Sandro de Silvestri
Via Mazzini 72
S. Colombano
Milano
Italy

Vern N. Smiley
United States of America
Office of Naval Research
223/231 Old Marylebone Road
London, NW1 5TH
England

Jeffrey I. Steinfeld
Massachusetts Institute of Technology
Department of Chemistry
Cambridge, Mass. 02133
USA

Stehlé
SOPRA
82 rue P. Brossolette
92250 La Garenne
France

Andreas Steudel
Institut A für Experimentalphysik
Technische Universität Hannover
Appelstraße 1
3 Hannover
Germany

Michael Stock
Fachbereich Physik der Universität
Kaiserslautern
6750 Kaiserslautern
Germany

M. Stuke
MPI für biophysikalische Chemie
Abt. Laserphysik
Postfach 968
34 Göttingen
Germany

Steven Suchard
US ERDA/NRA
M/S H 407
Washington, D.C. 20545
USA

Hideo Tashiro
Max-Planck-Institut für biophysika-
lische Chemie
Postfach 968
3400 Göttingen
Germany

Joel Tellinghuisen
Department of Chemistry
Vanderbilt University
Nashville, Tennessee 37235
USA

Herbert Walther
Sektion Physik der Universität
München
Coulombwall 1
8046 Garching
Germany

K.H. Welge
Universität Bielefeld
Fakultät für Physik
48 Bielefeld
Germany

Leroy E. Wilson
2932 LaVeta, N.E.
Albuquerque, N.M. 87110
USA

J.R. Wiesenfeld
Department of Chemistry
Cornell University
Ithaca, N.Y. 14853
USA

Jack Wilson
Laboratory for Laser Energetics
Universität of Rochester
Rochester N.Y. 14627
USA

Siegbert Witkowski
Projektgruppe für Laserforschung
der Max-Planck-Gesellschaft
8046 Garching
Germany

Klaus J. Witte
Projektgruppe für Laserforschung
der Max-Planck-Gesellschaft
8046 Garching
Germany

W.J. Wittemann
Physics Department
Twente University of Technology
P.O.Box 217
Enschede
Netherlands

Jürgen Wolfrum
Max-Planck-Institut für Strö-
mungsforschung
Böttingerstr. 6-8
3400 Göttingen
Germany

Walter G. Wrobel
Aschheimerstr. 16
8043 Unterföhring
Germany

Federico Zaraga
Istituto di Fisica del Polytecnico
Piazza Leonardo da Vinci 32
Milano
Italy

Werner Zapka
Kalklage 7
3400 Göttingen
Germany

Titles of Related Interest

Springer-Verlag
Berlin Heidelberg New York